ADVANCES IN ENVIRONMENTAL SCIENCE AND ENGINEERING

ADVANCES IN ENVIRONMENTAL SCIENCE AND ENGINEERING

MANUSCRIPT SUBMISSION The third annual volume contains invited articles from renowned scholars in their fields.
Unsolicited manuscripts are encouraged and we shall give them our most careful attention. Such articles should be sent directly to the editors.

ISSN 0141-8106

ADVANCES IN ENVIRONMENTAL SCIENCE AND ENGINEERING

Edited by

James R. Pfafflin and Edward N. Ziegler

VOLUME 3

GORDON AND BREACH SCIENCE PUBLISHERS
New York London Paris

Gordon and Breach Science Publishers, Inc.
One Park Avenue
New York, NY 10016

Gordon and Breach Science Publishers Ltd.
42 William IV Street
London WC2N 4DE

Gordon & Breach
7–9 rue Emile Dubois
Paris 75014
France

Library of Congress catalog card number 79-03896. ISBN 0 677 1576 6. Printed in Great Britain.

ACKNOWLEDGMENT

The editors wish to express their appreciation to Mrs. Katherine Jakobsen, Secretary to the Department of Chemical Engineering at the Polytechnic Institute of New York. Her invaluable assistance in the preparation of manuscripts and her attention to the small details were so important to this book.

CONTENTS

Preface xi

Contributors xiii

THE 1977 AMENDMENTS TO THE FEDERAL WATER POLLUTION CONTROL ACT, William Goldfarb 1

INTEGRATION OF PHYSICAL–CHEMICAL PROCESSES IN BIOLOGICAL WASTEWATER TREATMENT, Jatinder K. Bewtra 14

RECENT ADVANCES IN HYDRODYNAMIC TREATMENT FOR A BUOYANT SURFACE DISCHARGE, Kohji Muraoka and Keiji Nakatsuji 30

INFLUENCE OF FOREST VEGETATION AND AGRICULTURE ON THE ACIDITY OF FRESH WATER, I.Th. Rosenqvist 56

OXIDATION AND OXIDATION–INHIBITION OF SULFUR DIOXIDE, Elmar R. Altwicker 80

NITROGEN-DIOXIDE ABSORPTION IN EVAPORATING AND CONDENSING WATER DROPLETS, John P. Hermann and Michael J. Matteson 92

THE CHEMISTRY AND TRANSPORT OF METHANE AND CARBON MONOXIDE IN THE TROPOSPHERE, Leonard K. Peters and William L. Chameides 100

RURAL OZONE—SOURCES AND TRANSPORT, James J. B. Worth and Lyman A. Ripperton 150

INFLUENCE OF POLLUTION ON URBAN CLIMATOLOGY, Robert D. Bornstein and Tim Oke . . 171

ELECTROSTATIC PRECIPITATORS FOR PARTICULATE CONTROL, Peter B. Lederman and Paul L. Feldman . . . 203

Index 221

PREFACE

Wolman† has pointed out that "It is undoubtedly true that the intellectual forces which created a better life for man, and likewise violated some of the environment, are the same forces to which one must look for correcting past and avoiding future deteriorations". The successful search for new fundamental knowledge and rapid technological advances are weapons in the arsenal of the environmental worker, on whom the world may so heavily depend. The environmental scientist and engineer, indeed all professionals and concerned citizens, are themselves dependent on dissemination of new knowledge. The work presented here, and in succeeding years, is and will be but a small contribution to the overall international effort. We have endeavoured in the Advances Series to bring together the recent contributions of respected scholars and practitioners and make their work and views available to a wide multidisciplinary audience. The techniques of environmental science and engineering are of great significance in maintaining and enhancing the future quality of terrestial life. They must be applied to both prepare for and counteract the impact of industrially, domestically and naturally produced deleterious substances on the atmosphere, the bodies of water, and the terrain.

Much of the philosophy and scope of the Advances in based on the editors' earlier *Encyclopedia of Environmental Science and Engineering*. However, as the challenges to our ecosystem change, the procedures for analyzing and ameliorating the effect of both presently identified and emerging detrimental factors will require revision and expansion of our existing knowledge. It is thus logical that an "Advances Series" should occupy a humble but finite space within our globe, hopefully with a long half life before its ultimate passage to recyclable solid waste.

Progress in environmental science may be regarded as occurring on two fronts. The first is increasing depth of knowledge within a specific field. The second is concerned with greater understanding of interactions among a

† Dr. Abel Wolman (1976), Introduction to Encyclopedia of Environmental Science and Engineering, Gordon and Breach Science Publishers, Ed.: Pfafflin, J. R. and E. N. Ziegler.

number of related disciplines. The relationships may not always be readily apparent. It is our intention to strike a balance in presenting coverage of the latest developments in both approaches.

The Advisory Board and Authors, international in scope and outlook, composed of distinguished scholars and practitioners, represent a broad spectrum of disciplines and endeavours.

CONTRIBUTORS TO VOLUME 3

ELMAR R. ALTWICKER, Department of Chemical and Environmental Engineering, Rensselaer Polytechnic Institute, Troy, New York (Oxidation and Oxidation–Inhibition of Sulfur Dioxide)

JATINDER K. BEWTRA, Department of Civil Engineering, University of Windsor, Windsor, Ontario (Integration of Physical–Chemical Processes in Biological Wastewater Treatment)

ROBERT D. BORNSTEIN, Department of Meteorology, San Jose State University, San Jose, California (Influence of Pollution on Urban Climatology)

WILLIAM L. CHAMEIDES, Department of Physics and Astronomy, University of Florida, Gainesville, Florida (The Chemistry and Transport of Methane and Carbon Monoxide in the Troposphere)

PAUL L. FELDMAN, Research Cottrell, Inc., Bound Brook, New Jersey (Electrostatic Precipitators for Particulate Control)

WILLIAM GOLDFARB, Department of Environmental Resources, Rutgers—The State University, New Brunswick, New Jersey (The 1977 Amendments to the Federal Water Pollution Control Act)

JOHN P. HERMANN, School of Chemical Engineering, Georgia Institute of Technology, Atlanta, Georgia (Nitrogen-dioxide Absorption in Evaporating and Condensing Water Droplets)

PETER B. LEDERMAN, Research Cottrell, Inc., Bound Brook, New Jersey (Electrostatic Precipitators for Particulate Control)

MICHAEL J. MATTESON, School of Chemical Engineering, Georgia Institute of Technology, Atlanta, Georgia (Nitrogen-dioxide Absorption in Evaporating and Condensing Water Droplets)

KOHJI MURAOKA, National Institute for Environmental Studies, Ibaraki, Japan (Recent Advances in Hydrodynamic Treatment for a Buoyant Surface Discharge)

KEIJI NAKATSUJI, Department of Civil Engineering, Osaka University, Osaka, Japan (Recent Advances in Hydrodynamic Treatment for a Buoyant Surface Discharge)

TIM OKE, Department of Geography, University of British Columbia, Vancouver, British Columbia (Influence of Pollution on Urban Climatology)

LEONARD K. PETERS, Department of Chemical Engineering, University of Kentucky, Lexington, Kentucky (The Chemistry and Transport of Methane and Carbon Monoxide in the Troposphere)

LYMAN A. RIPPERTON, Research Triangle Institute, Research Triangle Park, North Carolina (Rural Ozone—Sources and Transport)

I. TH. ROSENQVIST, Institute of Geology, University of Oslo, Oslo, Norway (Influence of Forest Vegetation and Agriculture on the Acidity of Freshwater)

JAMES J. B. WORTH, Research Triangle Institute, Research Triangle Park, North Carolina (Rural Ozone—Sources and Transport)

THE 1977 AMENDMENTS TO THE FEDERAL WATER POLLUTION CONTROL ACT†

On December 28, 1977, President Carter signed into law the Clean Water Act of 1977,[2] which consists of an extensive set of amendments to the Federal Water Pollution Control Act—last significantly amended in 1972. In confusing fashion, the Clean Water Act of 1977 states that the Federal Water Pollution Control Act may also be referred to as the Clean Water Act. Nevertheless, its official title remains the Federal Water Pollution Control Act, and this is the title (abbreviated to F.W.P.C.A.) which will be utilized in this article. In order to avoid further confusion, the Clean Water Act of 1977 will be referred to as "the 1977 amendments."

The 1977 amendments modify nearly all major aspects of the F.W.P.C.A.

1 Industrial Point Sources

In 1972, the F.W.P.C.A. was amended to establish "a national goal that the discharge of pollutants into the navigable waters be eliminated by 1985."[3] An intermediate goal, the so-called "fishable/swimmable goal," was "that wherever attainable, an interim goal of water quality which provides for the protection and propagation of fish, shellfish, and wildlife and provides for recreation in and on the water be achieved by July 1, 1983." Closely related to these goals were the following requirements for industrial-point-source dischargers: (1) for existing sources, compliance by July 1, 1977, with effluent limitations based on Best Practicable Control Technology Currently Available (B.P.T.) or, if more stringent, with effluent limitations based on water-quality standards reflecting secondary contact uses (e.g., fishing); and by July 1, 1983, with effluent limitations based on Best Available Technology Economically Achievable (B.A.T.), or more stringent water-quality-based effluent limitations derived from water-quality standards reflecting fishable/swimmable uses, wherever attainable; and (2) for new sources,

† This article supplements "Federal Water Pollution Control Law Since 1972," which appeared in Volume 2 of *Advances in Environmental Science and Engineering*.

immediate compliance with promulgated standards of performance based on Best Available Demonstrated Control Technology (B.A.D.C.T). The 1972 amendments provided for two variances from technology-based effluent limitations: (1) for dischargers of heat, a variance predicated upon assuring "the protection and propagation of a balanced, indigenous population of shellfish, fish, and wildlife in and on the body of water into which the discharge is to be made;" and (2) in the event that more stringent water-quality-based effluent limitations were not required, a variance from effluent limitations based on B.A.T. for any discharger capable of showing that the modified requirement would "represent the maximum use of technology within the economic capability of the owner or operator" *and* "result in reasonable further progress toward the elimination of discharge of pollutants." The industrial-point-source requirements in the 1972 amendments are enforced through the National Pollutant Discharge Elimination System (N.P.D.E.S.) permit system, which incorporates the F.W.P.C.A.'s fundamental principle that any discharge in violation of the above-mentioned requirements is illegal.

With regard to industrial-point-source discharges, the 1977 amendments do not alter the basic philosophy and regulatory structure of the F.W.P.C.A. The 1985 "zero discharge" and 1983 "fishable/swimmable" goals have been retained. Furthermore, the N.P.D.E.S. permit program, with its investigation and enforcement components, have not been significantly altered. But even though the procedural aspects of the N.P.D.E.S. permit program are virtually untouched by the 1977 amendments, there are important changes in the substance of effluent limitations and compliance deadlines.

First, limited extensions of the July 1, 1977 deadline for compliance by industrial-point-source dischargers with B.P.T.-based effluent limitations are now available. The E.P.A. Administrator is authorized to allow extensions—until April 1, 1979 at the latest—where a discharger (1) has acted in good faith; (2) has made a serious commitment to achieving compliance; (3) will not, through obtaining an extension, cause the imposition of stricter controls on other sources; (4) has applied for a permit prior to December 31, 1974; and (5) has begun construction work on facilities necessary for compliance. This statutory extension mechanism strengthens and codifies E.P.A.'s administrative Enforcement Compliance Schedule Letter (E.C.S.L.) strategy. E.P.A. estimates that such extensions will be available to approximately ten percent of all major dischargers. Another type of extension is open to industrial-point-source dischargers which will agree to "tie in" to publicly owned treatment works, or which have already agreed to tie in but the municipal works is "presently unable to accept such discharge without construction." Extensions of this second variety can be granted until July 1, 1983 at the latest.

These new extensions of the 1977 deadline are both simple and narrow. However, the 1977 amendments also provide for extensions of the July 1, 1983 deadline for compliance with B.A.T.-based effluent limitations; and it is here that major substantive—although not structural or philosophical—changes are made in the F.W.P.C.A.

Under the 1972 amendments, there was no clear distinction among types of pollutants except for heat. Dischargers of all pollutants except for heat were subject to the 1977 and 1983 deadlines for compliance with effluent limitations based either on technology or, where more stringent, on water-quality standards. But with regard to toxic pollutants a good deal of confusion arose because of the existence of a separate, alternative approach requiring the Administrator (1) to publish a list of toxic pollutants; (2) to propose effluent standards for these pollutants based not on technology or water-quality standards but on toxicity, persistence, degradability and ecosystem effects; (3) to set a margin of safety; (4) to hold lengthy hearings; and (5) to enforce compliance with the standards within one year of promulgation. The unworkability of this approach convinced E.P.A. to ignore it, and administrative inaction stimulated a number of lawsuits against E.P.A. Finally, these lawsuits were settled, and the now famous "Consent Decree" confirmed E.P.A.'s decision to regulate toxic pollutants generally by requiring dischargers to comply with effluent limitations based on B.A.T. The Consent Decree also established timetables for E.P.A. to promulgate effluent limitations for 65 identified toxic pollutants applicable to 21 industrial subcategories. These timetables required all dischargers in these subcategories to meet their effluent limitations for toxic pollutants within three years of promulgation, or by July 1, 1983 at the latest. Thus, the Consent Decree preserved the F.W.P.C.A.'s basic mandate that all dischargers achieve B.A.T. by 1983.

The changes made by the 1977 amendments can best be understood as a codification and modification of the Consent Decree. There are now three types of pollutants other than heat: toxic pollutants, conventional pollutants and nonconventional pollutants. Toxic pollutants are those 65 materials which appear in the Consent Decree and any additional toxics which E.P.A. may list in the future. Dischargers of toxic pollutants much achieve B.A.T.-based effluent limitations by July 1, 1984 for the 65 Consent Decree toxics (a one-year postponement), or within three years of promulgation for newly listed toxics. But E.P.A. can require zero discharge or a one-year compliance period where necessary. There are no waivers or variances where toxic pollutants are concerned.

Conventional pollutants are defined as at least biochemical oxygen demand, suspended solids, fecal coliform and pH. For dischargers of these pollutants, effluent limitations based on Best Conventional Pollutant Control Technology (B.C.T.) must be achieved by July 1, 1984. B.C.T. includes a new "cost test" whereby E.P.A. must consider "the reasonableness of the relationship between the costs of attaining a reduction in effluents and the effluent reduction benefits derived . . . ." The test is also to consider the cost to a publicly owned treatment works of achieving proportional effluent reductions. If the industrial cost is less than the proportional municipal cost, the industrial discharger will generally be required to upgrade treatment. This cost test for B.C.T. is far from clear, and will be difficult to apply in practice. However, the Conference Committee believed that B.C.T. will most often fall between B.P.T. and B.A.T. Dischargers of conventional pollutants will not be required to meet B.A.T.-based effluent limitations unless meeting water-quality standards necessitates effluent limitations more stringent than those based on B.C.T. Because of the general relaxation of treatment standards for conventional pollutants, there is no longer a variance for individual dischargers of conventionals.

Nonconventional pollutants are all those pollutants which are neither toxic pollutants nor conventional pollutants. Until a pollutant is added to E.P.A.'s toxic list it remains a nonconventional pollutant. Dischargers of nonconventional pollutants must meet B.A.T.-based effluent limitations between 1984 and 1987—a possible four-year postponement of the original Phase-II deadlines. Furthermore, there are two concurrent, alternative waiver opportunities for dischargers of nonconventional pollutants. First, there is the modification which "will represent the maximum use of technology within the economic capability of the owner and operator" and "will result in reasonable further progress toward the elimination of the discharge of pollutants." A second waiver provision has been added by the 1977 amendments. The Administrator, with the consent of "the State," may grant a waiver if the substitute effluent limitation (1) will be at least as strict as one based on B.P.T.; (2) will not result in additional requirements being placed on another point source; and (3) will not interfere with the attainment or maintenance of the fishable/swimmable goal or endanger human health. The 1977 amendments make no change in the F.W.P.C.A.'s treatment of heat but it is unclear whether the compliance date for dischargers of heat has been extended to 1984. New sources of pollution are still called upon to meet standards of performance based on Best Available Demonstrated Control Technology.

In conclusion, the 1977 amendments disrupt the balance between the goals and requirements of the F.W.P.C.A. The Phase-II compliance dates now exceed the 1983 goal of fishable/swimmable waterways wherever attainable.

It remains to be seen what effect this incongruity will have on the administration of the F.W.P.C.A.

2 Publicly Owned Treatment Works and Industrial Tie-ins

2.1 Compliance-date Extensions

With only approximately one-third of all municipal dischargers in compliance with the 1977 secondary-treatment deadline, and with courts refusing to grant extensions despite delays in federal funding, it was predictable that Congress itself would provide for extensions.

E.P.A. or an N.P.D.E.S. state may grant extensions—until July 1, 1983 at the latest—to publicly owned treatment works which have not achieved compliance as a result of construction delays or lack of promised federal funds. If an extension is granted, E.P.A. or a state will issue a new N.P.D.E.S. permit containing an appropriate compliance schedule. Industrial-point-source dischargers dependent upon treatment-plant construction in order to tie into the plant may also be granted extensions of their compliance dates until the completion of plant construction.

There is also a modification provision for publicly owned treatment works which discharge into marine waters. E.P.A., with the concurrence of the state of location, may modify the secondary-treatment requirement for "the discharge of any pollutant in an existing discharge from a publicly owned treatment plant into marine waters" where the modification will not exacerbate existing pollution problems or generate new ones. The term "marine waters" includes not only the territorial sea and contiguous zone but also "saline estuarine waters where there is strong tidal movement and other hydrological and geological characteristics" which will ensure dilution and dissipation. The Conference Report on the 1977 amendments indicates that marine-discharge modifications will be reevaluated at five-year intervals, and that no modification will be granted for discharges of persistent pollutants such as PCBs. E.P.A. has announced that such modifications will be available only to approximately 50 California communities.

2.2 Funding Authorization, Eligibility and Administration

The 1977 amendments authorize for the construction-grants program \$4.5 billion for fiscal year 1978, and \$5 billion for fiscal years 1979 through 1982. However, these authorizations are subject to appropriations acts rather than being available for immediate, binding allotment as was the case under the 1972 amendments. In the words of the Conference Committee, "an allotment is no longer a commitment to fund because, although contract authority is still theoretically available, it no longer has any impact on the availability of funds." Nevertheless, Congressional appropriations committees can make advanced appropriations for the construction-grants program.

Treatment works for the control of pollutant discharges from separate storm-sewer systems are no longer grant-eligible; but projects to separate storm and sanitary sewers are unaffected if they do not include independent treatment facilities for storm water. Land necessary for the storage of treated wastewater prior to land application and land for sludge treatment are now explicitly eligible for funding.

Attempts by E.P.A. and environmentalists to limit the reserve capacity of publicly owned treatment works and to eliminate collector systems and major sewer rehabilitation from grant eligibility were frustrated. The provision of the 1977 amendments regarding reserve capacity is an ambiguous clause which vests complete discretion in E.P.A., with no firm guidelines. Not only are collector systems still eligible for funding, but "if the Administrator uses population density as a test for determining the eligibility of a collector sewer for assistance it shall be only for the purpose of evaluating alternatives and determining the needs for such system in relation to ground or surface water quality impact." Moreover, each state must use at least 25% of its alloted construction-grant funds for major sewer rehabilitation, new collector sewers, new interceptors and correction of combined sewer overflows if projects in these categories are on the state's priority list.

The 1977 amendments encourage and authorize E.P.A. to delegate management of the construction-grants program to the states. E.P.A. can reserve up to 2% of a state's allotment, but not less than $400,000, for state administration of the program. State priority lists are binding on E.P.A. unless "the Administrator, after a public hearing, determines that a specific project will not result in compliance with the enforceable requirements" of the F.W.P.C.A.

2.3 Innovative and Alternative Technology and Management

Environmentalists were successful in obtaining a number of incentives for the development and application of innovative and alternative waste-treatment technology and management techniques. No construction grant can be made unless the applicant demonstrates that "innovative and alternative wastewater treatment processes and techniques which provide for the reclaiming and reuse of water, otherwise eliminate the discharge of pollutants, and utilize recycling techniques, land treatment, new or improved methods of waste treatment management for municipal and industrial waste and the confined disposal of pollutants . . . have been fully studied by the applicant taking into account . . . the more efficient use of energy resources." Current advanced-treatment technologies do not qualify as innovative and alternative methods.

By July 1, 1979, E.P.A. must publish guidelines for the identification and evaluation of innovative and alternative treatment processes and techniques. Conforming projects may be funded, even though they may cost up to 15% more than the most cost-effective alternative. Moreover, the grant share for such facilities and management techniques is 85%, instead of 75% for conventional treatment works, and extra grant monies will be available for modification or replacement of the funded facilities if they do not meet design performance specifications (except in cases of negligence). In fiscal years 1981 and 1982, at least 3% of each state's allotment must be expended only for increasing construction grants from 75% to 85%.

Beginning with fiscal year 1980, innovative technology must be used in treatment works at federal facilities unless the system is more than 15% more expensive than the most cost-effective alternative, or unless E.P.A. waives the requirement in the public interest. States may modify their priority lists so as to accord a higher priority to projects using innovative technology, but such a reorientation is not mandatory. Finally, private firms are rewarded for adopting innovative technology. Industries proposing to install innovative production processes or control techniques which would achieve "an effluent reduction significantly greater than that required by the limitation otherwise applicable," or which will meet the applicable effluent limitation at significantly lower cost than existing systems, may be granted extensions of the B.A.T.-compliance dates until 1987 if the innovative systems possess the potential for industry-wide application.

Federal funding is now available for the construction of privately owned treatment works serving one or more residences or small commercial establishments. A public body must apply for the grant, certify that public ownership is infeasible and guarantee proper operation and maintenance through a system of user charges. Such private systems must be cost-effective and must provide for cost recovery from the commercial establishments serviced. The Conference Report makes it clear that grants for privately owned treatment works will not be made for individual septic tanks or secondary package plants, but will be restricted to "alternative or unconventional treatment works for individual residences or clusters of residences."

Under the so-called "Rural Set-aside" clause, E.P.A. will set aside 4% of the allotment of each state with a rural population of 25% or more.

(There is also a discretionary set-aside for other states upon request by their Governors.) These earmarked funds are available only for alternatives to conventional treatment works for municipalities with populations of 3500 or less, or for "the highly dispersed sections of larger municipalities."

2.4 User Charges and Industrial Tie-Ins

As expected, existing *ad valorem* user charges are now allowed, but only with regard to residential and small nonresidential users of publicly owned treatment works. In order to qualify as a user charge, the *ad valorem* levy must be dedicated to the operation and maintenance of the treatment works, charges must be proportionately distributed between classes and industrial-cost recovery must be in operation. A nonmetering residential-user charge system—including an *ad valorem* system—may be approved if revenues are dedicated, users are notified of their proportionate charges for waste-treatment services and proportionality is maintained both within and between classes. However, metering is still required for industrial users.

There is a moratorium on industrial-cost recovery until July 1, 1979, during which time E.P.A. will study the impacts of industrial-cost recovery and report to Congress. If Congress does not amend the F.W.P.C.A. in this area, charges for the moratorium period will be collected as prorated over the remaining life of the treatment works. Small indirect dischargers with flow rates of 25,000 gpd or less of sanitary wastes are exempted from industrial-cost recovery unless the discharges upset the treatment process or contaminate the sludge. Industrial-cost recovery may now be based on use of the entire system instead of only the treatment plant, and reductions may be made for flow diminution and water conservation.

The 1977 amendments clarify the requirements with regard to toxic pretreatment standards. E.P.A. will establish national pretreatment standards for toxic pollutants based on B.A.T. or any more stringent limitations set under section 307. Then, in applying these pretreatment standards through its pretreatment program, the owner or operator of a publicly owned treatment works can modify the requirements applicable to tie-in sources to reflect the degree of reduction of that pollutant achieved by the treatment works. The combination of pretreatment and treatment by the municipal treatment works shall attain at least that level of treatment which would be required if the tie-in source were making a direct discharge to a waterway. Finally, no modification of pretreatment standards can be granted if it will result in sludge contamination.

As for sludge, E.P.A. will promulgate by July 1, 1979 regulations containing guidelines for sludge disposal or utilization. Decisions about sludge disposal or utilization are to be made at the local level, but violation of E.P.A. sludge regulations is unlawful.

In sum, the political trade-off which appears to have been made with regard to publicly owned treatment works is a new emphasis on innovative and alternative technology in exchange for maintaining the *status quo* as to reserve and collector capacity. Only time will tell whether the vision of innovative and alternative technology is illusory in the light of the sewage-treatment community's dedication to conventional technologies.

3 Nonpoint-Source Control

According to a recent Comptroller General's Report to the Congress, "nonpoint" sources of water pollution (sediment, acid mine drainage, pesticides and other pollutants carried into streams by runoff from rainstorms)

> currently produce more than half of the pollutants entering the Nation's waterways. GAO reviewed overall efforts to control nonpoint sources of pollution and concluded that progress has been minimal. If not controlled, nonpoint pollution will prevent attainment of national water quality goals and will continue to grow in significance as "point" sources of pollution such as factories and muncipal waste treatment plants are brought under control.[4]

Thus, pollution from nonpoint sources has become our most difficult water-pollution-control problem.

In the 1972 amendments, Congress concentrated on the regulation of point sources of pollution. E.P.A. was given no direct authority to regulate nonpoint sources. (Specifically, nonpoint source dischargers were not required to possess N.P.D.E.S. permits; and the construction grants program did not apply to nonpoint-source control mechanisms.) Under section 208 of the F.W.P.C.A., states and designated regional planning agencies were expected to develop water-quality-management plans (1) identifying water-quality problems caused by nonpoint sources, (2) establishing Best Management Practices (B.M.P.) for such sources and (3) creating implementable regulatory programs to control nonpoint-source pollution. However, the 1972 amendments did not contain sanctions adequate to ensure cogent nonpoint-source planning and effective implementation of approved plans.[5] Consequently, many water-quality-management plans submitted for state certification or E.P.A. approval are defective with regard to the control of nonpoint-source pollution.

The 1977 amendments do not significantly change the legal structure of nonpoint-source control. Federal law still does not require federal or state permits for most nonpoint sources; and the construction-grants program, although expanded in other areas, is still limited to point sources or groups of point sources. Moreover, no new sanctions have been adopted to penalize defective planning or failure to implement water-quality-management plans.

Nevertheless, the 1977 amendments do evidence an incipient trend toward requiring permits for nonpoint sources and providing federal subsidies for nonpoint-source control.

E.P.A. is authorized to publish regulations amending effluent-limitation guidelines for point sources, to include B.M.P. for ancillary industrial activities which may contribute significant amounts of toxic or hazardous pollutants to navigable waterways. The intent of this provision is not to involve E.P.A. in actual plant process-design and operating decisions, but to control the runoff of toxic and hazardous materials from industrial sites resulting from poor housekeeping practices such as plant-site runoff, spillage or leaks, sludge or waste disposal and drainage from raw-material storage. Once promulgated as regulations, B.M.P. must be included in N.P.D.E.S. permits for relevant point sources.

A major federal subsidy for nonpoint-source control has been added in the form of agricultural cost sharing. The Secretary of Agriculture, acting through the Soil Conservation Service (S.C.S.), is authorized to enter into contracts of between five and ten years with "owners and operators having control of rural land" for the purpose of installing and maintaining B.M.P. in those areas where water-quality-management plans are in effect and the control practices specified in the contracts have been certified by the appropriate management agencies as being consistent with their plans. If a particular landowner or operator does not carry out the terms of his contract "substantially in accordance with the schedule contained therein," the contract can either be terminated and the owner compelled to forfeit future payments and refund prior payments, with interest; or the S.C.S. can demand refunds and adjustments instead of terminating the contract. Forfeit and refund is required where the land is transferred, unless the transferee agrees to be bound by the contract. S.C.S. is authorized to provide technical assistance and up to 50% of the total cost of B.M.P., except that the matching share may be increased where "(1) the main benefits to be derived from the measure are related to improving offsite water quality, and (2) the matching share requirement would place a burden on the landowner which would probably prevent him from participating in the program." The 1977 amendments authorize $200 million for fiscal year 1979 and $400 million for fiscal year 1980 to fund agricultural cost sharing.

The agricultural-cost-sharing provisions of the 1977 amendments emphasize S.C.S.'s responsibility to give priority "to those areas and sources that have the most significant effect on water quality." According to the Conference Report, the function of the cost-sharing program is to reduce nonpoint source pollution through financial assistance for only those soil conservation practices which improve water quality. It is not intended to be a copy or extension of existing soil conservation programs in the Department of Agriculture, and should not finance production-oriented practices except as an incidental or indirect result.

However, it is not clear why a farmer or rancher would participate in the program at less than 100% funding—if then—in the absence of a concurrent standard-setting and enforcement program to control agricultural runoff. It is also conceivable that exising rivalries within the Department of Agriculture will deflect the program away from water-quality improvement and toward increasing agricultural production.

The 1977 amendments change the status of irrigation return flows from agricultural land. Whereas under the 1972 amendments irrigation return flows were classified as point sources and thus subject to the N.P.D.E.S. permit program, they are now explicitly designated as nonpoint sources to be included in water-quality-management planning.

Congress has authorized $150 million per fiscal year through 1980 for the water-quality-management planning program, and the federal share of most future planning costs incurred by designated agencies has been set at 75%. As for state planning areas, E.P.A. is authorized to utilize a portion of a state's construction-grants allotment—over and above the 2% state administration fund—for the purpose of making grants to state agencies for "administering a statewide waste treatment management planning program." A sum of $100 million per fiscal year through 1980 is authorized for federal interagency agreements to utilize other federal laws and programs for the implementation of water-quality-management plans.

The planning period has been increased by one year to a more realistic three years; and water-quality-management plans must now contain, in addition to irrigation-return-flow components, planning for recreation and other multiple-use benefits from improved water quality and land associated with publicly owned treatment works.

In summary, the 1977 amendments have made welcome but incremental changes in the F.W.P.C.A.'s planning sections. Unless E.P.A. develops meaningful administrative sanctions for defective planning or nonimplementation, Congress may be reluctant actually to appropriate money for future water-quality-management planning.

4 The Section-404 Program

After much clamor and controversy, the section-404 permit program regulating discharges of dredged or fill material into waterways or wetlands was continued in modified form by the 1977 amendments.

The Corps of Engineers (C.O.E.) is still authorized to grant individual permits for discharges of dredged or fill material, but it may now issue general permits on a state, regional or nationwide basis for activities which are similar in nature and environmentally insignificant. General permits must conform to guidelines promulgated by E.P.A., and these permits may be for terms of five years or less, subject to revocation for cause.

The following discharges of dredged and fill material are specifically exempted from the requirement of obtaining a permit: (1) those from normal farming, silviculture and ranching activities; (2) those from the maintenance or repair of existing structures; (3) those from the construction or maintenance of farm or stock ponds or irrigation ditches, or the maintenance of drainage ditches; (4) those from the construction of sedimentation basins on a construction site which does not include placement of fill material in waterways and wetlands; (5) those from the construction or maintenance of farm, forest or mining roads in accordance with B.M.P.; (6) those from any activity regulated by a state under a federally approved system of B.M.P. applied on a statewide basis; and (7) those from any fully federal project authorized by the entire Congress, if an environmental-impact statement which adequately addresses the project's potential discharge of dredged or fill material has been submitted to Congress prior to discharge and prior to Congress' authorization or appropriation of funds for the project. However, these exemptions are themselves subject to two important exceptions. Permits are required even for activities in the above categories where (1) such discharges are "incidental to any activity having as its purpose bringing an area of the navigable waters into a use to which it was not previously subject, where the flow or circulation of navigable waters may be impaired or the reach of such waters be reduced"; or (2) such discharges contain toxic pollutants listed under section 307 of the F.W.P.C.A.

States can elect to administer their own permit programs for the regulation of discharges of dredged or fill materials into the navigable waters other than traditionally navigable waters and adjacent wetlands. (In other words, state permit programs may only apply to certain freshwaters and adjacent wetlands.) Once a state program has been approved, the federal program is thereby suspended in appropriate areas of that state. The process of state-program approval and revocation is virtually identical to that involving state administration of the N.P.D.E.S. permit program. It is important to note, however, that state permit programs for dredged and fill materials must be approved not by C.O.E. but by E.P.A. in consultation with C.O.E. and the Fish and Wildlife Service. Consequently, close cooperation and coordination is as necessary to the success of state-program development as it is to the viability of section 404 in general. (For example, E.P.A. formulates the guidelines which C.O.E. applies in the permit program.) Unless E.P.A. and C.O.E. can transcend historical differences, Congress may not have heard the last of section 404.

REFERENCES

1. W. Goldfarb, "Federal Water Pollution Control Law Since 1972," in *Advances in Environmental Science and Engineering*, Vol. 2, edited by J. R. Pfafflin and E. N. Ziegler (Gordon and Breach, New York, 1979).
2. *Federal Water Pollution Control Act, as Amended by Clean Water Act of 1977*, 33 U.S.C. 1251, *et seq.*
3. W. Goldfarb, "Better Than Best: A Crosscurrent in the Federal Water Pollution Control Act Amendments of 1972," *Land and Water Law Rev.*, **11**, No. 1 (1976).
4. Comptroller General of the U.S.A., *National Water Quality Goals Cannot Be Attained without More Attention to Pollution from Diffused or "Nonpoint" Sources*, (December 20, 1977).
5. W. Goldfarb, "Water Quality Management Planning: The Fate of 208," *Toledo Law Rev.*, **8**, 105 (1976).

William Goldfarb
Rutgers University

INTEGRATION OF PHYSICAL–CHEMICAL PROCESSES IN BIOLOGICAL WASTEWATER TREATMENT

Introduction

The present conventional wastewater-treatment practices have developed in different stages over a long period of time. The first stage of development was preliminary treatment for the removal of inert suspended particles, in order to protect pumps and subsequent treatment units. This was followed by primary treatment for the separation of suspended solids (SS) along with reduction in the biochemical oxygen demand (BOD) of effluent. Subsequently, secondary treatment was added to reduce the BOD further and to separate the suspended solids before the disposal of the final effluent. Traditionally, the secondary treatment has been in the form of biological flocculation and oxidation of organic matter.

Recent years have seen the introduction of several new processes to improve the scope, capacity and efficiency of wastewater-treatment plants, in order to meet the more stringent effluent standards. This has involved both the development of new technology and the adoption of the known technology, after making certain modifications, in existing treatment plants. The most important development in wastewater-treatment technology in the last decade has been the addition of physical–chemical processes. When these physical–chemical processes are used in addition to the biological treatment to remove pollutants which are not adequately removed otherwise, it is considered to be tertiary treatment or advanced wastewater treatment. The two most common physical–chemical unit operations adopted in recent years are chemical coagulation and flocculation for the removal of phosphates and chlorination or ozonation for disinfection.

Some design engineers have considered the exclusive use of physical–chemical processes as a possible replacement for the conventional biological treatment. It is proving impractical, especially for domestic wastewaters and certain industrial wastewaters; at the same time, however, the importance of physical–chemical processes is being recognized. The zero-discharge goal is a clear indication of a fast move towards the integration of practicable physical–chemical unit operations into conventional biological-treatment plants.

With the introduction of physical–chemical processes, a new design approach has developed: specific units are designed for the removal of specific pollutants and are then integrated together on an optimal basis. The scope and limitations of integrating physical–chemical unit operations into conventional biological-treatment plants are considered and assessed in this article. Nonconventional methods or quasisecondary treatments such as lagoons, land treatment, etc., although equally significant, are not considered; likewise, discussion of direct physical–chemical wastewater treatment (a treatment system depending only on physical and chemical processes) has deliberately been omitted.

Sludge treatment is considered to be an integral part of wastewater treatment because, without proper sludge treatment, wastewater treatment becomes merely the removal of pollutants from the liquid to the sludge phase, thereby producing no improvement in the "total environment." In many instances, the disposal of residue becomes the major factor in the selection of treatment processes.

Wastewater Characteristics

This discussion will be limited to domestic wastewaters or industrial wastewater with similar characteristics. The important characteristics in the design and operation of such treatment plants are:

1) the flow rate, Q, and its variation;

2) nutrients
   a) phosphorus, P
   b) nitrogen, N;

3) decomposable organic matter causing biochemical oxygen demand, BOD
   a) volatile suspended solids, VSS
   b) volatile dissolved solids;

4) fixed solids
   a) total residue
   b) suspended residue;

5) coliform bacteria.

Typical domestic-wastewater characteristics and the desired effluent-quality requirements used for discussion in this article are summarized in Table I.

TABLE I

Wastewater characteristics and effluent-quality requirements

Parameter	Typical values for domestic wastewater[a]	Effluent-quality requirements[a]
Biochemical oxygen demand, BOD (5-day, 20°C)	Total BOD = 200 mg/liter Soluble BOD = 60 mg/liter	From less than 3 to less than 20 mg/liter
Chemical oxygen demand, COD	Total COD = 450 mg/liter Soluble COD = 250 mg/liter	
Suspended solids, SS	Total SS = 250 mg/liter VSS = 175 mg/liter	From less than 10 to less than 20 mg/liter
Phosphorus as P	Total P = 15 mg/liter Soluble P = 12 mg/liter	Less than 1.0 mg/liter
Nitrogen as N	Total N = 40 mg/liter NH_3–N = 25 mg/liter	Total N less than 4.0 mg/liter; unoxidized N less than 2.0 mg/liter
Coliform	50×10^6 MPN/100 ml	From less than 2 MPN/100 ml to less than 5000/100 ml

[a] The figures given correspond to the following conditions: pH, 7.2–7.7; alkalinity, 200 mg/liter as $CaCO_3$; grease, 100 mg/liter; total dissolved solids, 700 mg/liter; total organic carbon, 200 mg/liter.

It is assumed that the suspended solids contain about 65% BOD, 25% P and 30% N. It will be seen later that normal variations in waste characteristics can be accommodated in the integration of the various unit operations.

Biological Treatments

The biological-treatment processes considered in this discussion are various activated sludge processes for the removal of dissolved organic matter, various nitrification–denitrification processes for nitrogen removal and anaerobic sludge digestion. These processes have been discussed in detail elsewhere.[1] The major advantages of biological wastewater treatment are (1) the complete oxidation of a part of the organic matter to a more stable form; (2) the smaller quantity of sludge produced for handling; (3) that it is the most economical in removing large quantities of biodegradable organic matter; and (4) that, being a natural degradation process, it leads to less environmental damage. Some of this biodegradable matter cannot be removed by any of the physical–chemical processes.

Physical–Chemical Processes

The physical–chemical unit operations considered are:

1) chemical coagulation, flocculation and sedimentation (various chemicals, e.g., lime, alum, and ferric salts, with and without polymers, have been used successfully);

2) ammonia stripping;
3) filtration;
4) activated carbon;
5) chlorination or ozonation for disinfection and chemical oxidation.

Unit operations such as ion-exchange, the reverse-osmosis process and distillation have deliberately been omitted, because they are still at the development stage, and have not been adopted on both an economical and a technological basis. The advantages of physical–chemical processes are the reduced land-area requirement, higher efficiency, more flexibility and consistency in plant performance and greater resistance to shock toxic and other chemical substances.[2]

Sludge Treatment and Disposal

The methods considered are sludge thickening, anaerobic digestion and sludge dewatering. It is assumed that the sludge, with or without treatment, will be disposed of on land.

Effluent Disposal

The following methods are considered for ultimate disposal of the effluent:

1) in natural water bodies: lakes, rivers, the ocean;
2) on land;
3) underground;
4) reuse/recycle.

The layout of a complete treatment system is shown in Figure 1. The units are placed in the sequence in which they normally appear in treatment plants, pilot plants or laboratory studies. The switching points indicate how different unit operations may be added or deleted in any scheme under consideration.

Selection of a Treatment Process

In choosing a treatment process, the following factors often dictate the decision:

a) the nature of the wastewater;
b) effluent quality and use;
c) the quality and quantity of the sludge; recovery and reuse of chemicals;
d) cost; economic feasibility;

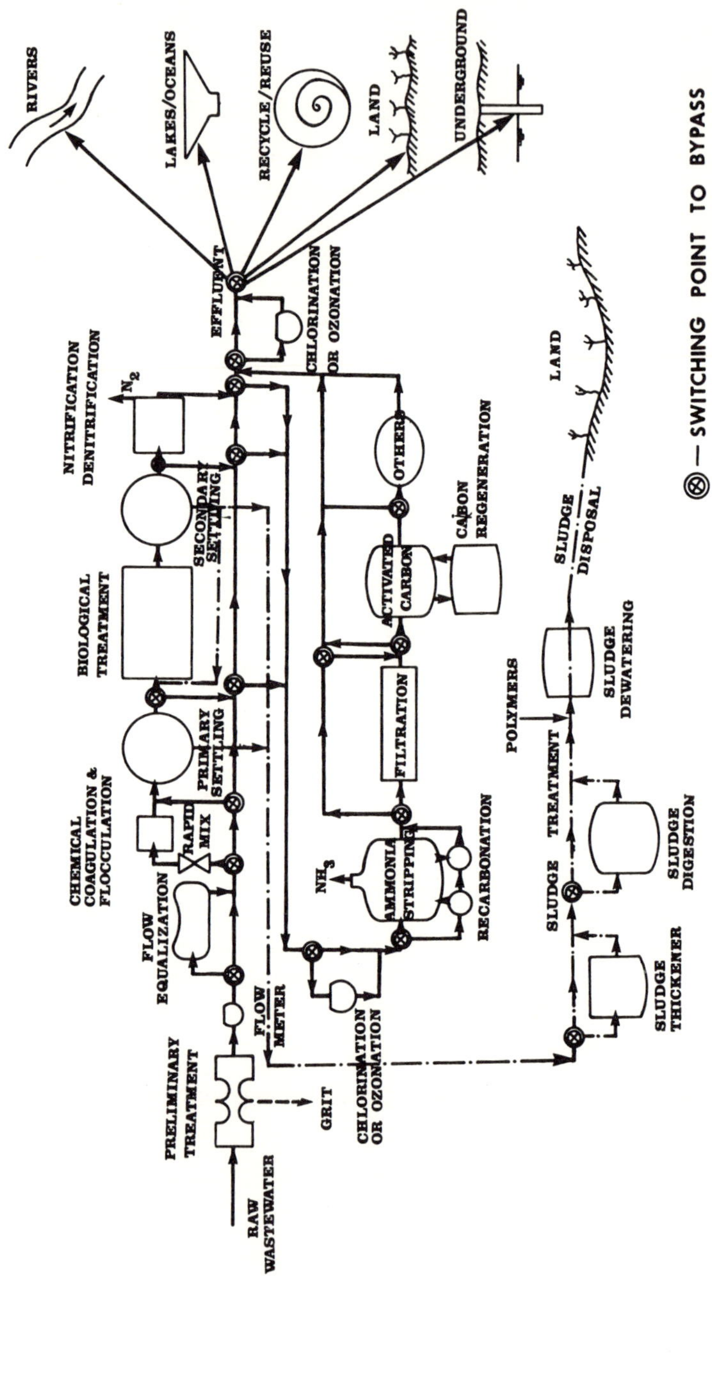

Figure 1 Wastewater treatment units.

e) energy requirements;
f) land requirement and available sites;
g) flexibility and reliability;
h) process compatibility;
i) operational and maintenance problems;
j) total environmental impact on water, air and land.

The stability of the process is very important: if the process is sensitive to wastewater characteristics, it may become necessary to choose a less economical but more stable process.

Economic Analysis

Topping[3] has developed cost relationships for different treatment-unit operations, with design capacities Q greater than 1.0 US MGD (0.0438 m^3/s), and these are summarized in Table II. The cost data are based on an E.P.A.

TABLE II

Cost relationships and land requirement for different unit operations[3]

Unit operation	Capital cost	Operation + maintenance cost, cents/1000 US gal	Land requirement,[a] acres
Primary treatment	$0.656Q^{0.673}$	$7.87Q^{-0.176}$	$0.230 + 0.088Q$
Chemical coagulation	$0.005Q^{0.680}$	$1.80Q^{-0.240}$	No land required
Activated sludge including primary treatment	$0.950Q^{0.778}$	$14.80Q^{-0.241}$	$2.212 + 2.438Q$
Nitrification–denitrification	$0.760Q^{0.700}$	$22.00Q^{-0.251}$	$0.622 + 0.4887Q$
Filtration	$0.140Q^{0.638}$	$7.48Q^{-0.294}$	$0.024 + 0.028Q$
Activated carbon	$0.170Q^{0.470}$	$5.24Q^{-0.154}$	$0.024 + 0.028Q$
Ozonation	$0.250Q^{0.730}$	$16.98Q^{-0.251}$	$-0.17 + 0.021Q$
Chlorination	$0.025Q^{0.674}$	$1.46Q^{-0.098}$	$-0.17 + 0.021Q$

[a] Neglect negative values; Q is in US MGD; 1000 US gal = 3.78 m^3; 1 US MGD = 0.0438 m^3/s; 1 acre = 0.4 ha.

wastewater-treatment-plant index of 212.15. These plants are expected to give the performance shown in Table III. The cost figures include the cost of sludge handling but not the cost of land, because it is assumed that the salvage value will at least remain equal to the prevailing market value. For any sequence of treatment-plant processes, an economic analysis for the overall cost of the treatment system (capital + operating + maintenance) should be performed, in accordance with standard practices.

TABLE III

Average performances and energy requirements for various treatment processes

Process train	Treatment processes[a]	Average residual concentration: SS, mg/liter	BOD, mg/liter	P, mg/liter	N, mg/liter	Coliform,[b] MPN/100 ml	Energy requirement for 100 MGD (4.38 m^3/s), kW/h/day
I	Preliminary treatment	250	200	15	40	5×10^7	250
II	PS	165	130	12	36	10^7	1,000
III	PS + CP	20	90	2.0	28	10^6	34,500
IV	PS + AS	60	30	9.0	30	10^6	93,000
V	PS + CP + AS	20	20	1.5	25	2×10^5	126,000
VI	PS + AS + Cl	60	25	9.0	30	5×10^3	93,500
VII	PS + CP + AS + Cl	20	18	1.5	25	10^3	126,500
VIII	PS + AS + ND	60	26	9.0	3.2		207,500
IX	PS + CP + AS + ND	20	16	1.5	3.2		241,000
X	PS + AS + ND + Cl	60	25	9.0	3.0		208,000
XI	PS + CP + AS + ND + Cl	20	15	1.5	3.0		241,500
XII	PS + AS + Cl + AM	60	25	9.0	3.0		359,000
XIII	PS + CP + AS + Cl + AM	20	18	1.5	3.0		392,000
XIV	PS + AS + ND + Cl + F	10	10	8.0	2.8		585,000
XV	PS + CP + AS + ND + Cl + F	5	6	1.2	2.8		618,500
XVI	PS + AS + Cl + AM + F	10	10	8.0	2.8		736,500
XVII	PS + CP + AS + Cl + AM + F	5	6	1.0	2.8		770,000
XVIII	PS + AS + ND + Cl + F + AC	8	2	7.0	2.4	2	802,500
XIX	PS + CP + AS + ND + Cl + F + AC	4	2	0.8	2.4	2	836,000
XX	PS + AS + Cl + AM + F + AC	8	2	7.0	2.4	2	954,000
XXI	PS + CP + AS + AM + Cl + F + AC	4	2	0.8	2.4	2	987,000

[a] Notation: PS, primary sedimentation; CP, chemical precipitation; AS, activated sludge; Cl, chlorination/ozonation; ND, nitrification–denitrification; AM, ammonia stripping; F, filtration; AC, activated carbon.

[b] Depends on the chlorine/ozone dose and the contact time.

Land Requirement

The land required in acres for each unit operation, as reported by Topping,[3] is given in Table II. This information becomes very significant in site selection when only a limited space is available at a premium price.

Variation in Flow Rate and Load

It has been recognized for a long time that both the flow rate and polutant loadings (flow × concentration) in a wastewater-treatment plant vary diurnally and daily. With stringent requirements on effluent quality, it has become important to take these variations into consideration when designing different components of wastewater-treatment plants. Young *et al.*[4] studied

TABLE IV

Summary of peaking ratios for Ames, Iowa[4]

Parameter	Peak month/av. annual	Peak day/av. annual	Peak 8 h/av. annual	Peak 4 h/av. annual
Flow	1.35	1.91	2.29	2.37
BOD load	1.30	1.59	2.23	2.39
SS load	1.37	2.28	3.05	2.27
NH_3–N load	1.25	1.51	2.11	2.27
PO_4 load	1.27	1.57	2.20	2.36

flow and constituent load variations for Ames, Iowa over a 23-year period of record and have reported the peaking ratios for flow and load shown in Table IV. They have further demonstrated how these variations can influence the performance of different treatment-plant components and have suggested design criteria for various units.

Reliability

Normally, the performance of a wastewater-treatment plant is reported in terms of daily, weekly or even monthly average values for the concentration and quantity of various substances in the effluent. These values alone provide no information on the statistical distribution of extreme values and thus on the ability of a plant to perform continuously within the established quality requirements. The reliability of any process can be described by its average performance and the standard deviation. According to Dean and Forsythe,[5] metal concentrations in sludges and the usual measurements of pollutants in wastewater effluents all have a log-normal distribution, i.e., logarithms of the values are normally distributed. Consequently, they have reported the effluent concentrations in terms of median values and a spread factor (the antilog of the standard deviation of the logarithms of the data points). For South Lake Tahoe treatment-plant effluent, subjected to conventional biological treatment followed by chemical treatment, carbon filters and chlorination, the spread factor values in Table V have been reported for different parameters.

TABLE V

Spread factors for South Lake Tahoe treatment-plant effluent

Parameter	Spread factor
BOD, mg/liter	2.05
COD, mg/liter	1.57
Turbidity, JTU	1.94
P, mg/liter	2.11
Coliform, MPN/100 ml	13.0

This clearly indicates that, if strict restrictions on effluent quality are to be imposed, information on the reliability of each unit operation is as important as its average performance. It is doubtful that wastewater-treatment plants that operate successfully for only part of the time will be acceptable in the future.

Energy Requirements

The energy crisis in recent years has made design engineers realize the importance of considering the energy requirements of the various wastewater-treatment-plant components in the course of process selection. Hagan and Roberts[6] have calculated the energy requirements (Table III) for a wastewater-treatment plant of capacity 100 US MGD (4.38 m^3/s) providing secondary and/or tertiary treatment and operating with the removal efficiencies shown in Table III. These figures for the energy requirement include values for sludge digestion followed by landfill disposal of digested sludge, manufacture and transport of chemicals, construction of facilities, etc.

Similar analysis for the energy requirements in Canada for conventional and advanced wastewater treatments were carried out by Zarnett[7]; his values

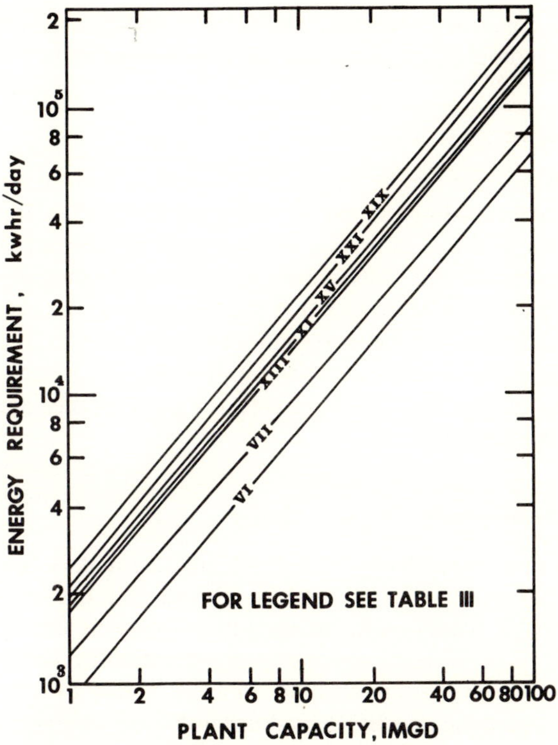

Figure 2 Energy requirements for various unit processes.

for plants of capacity 1–100 IMGD (0.053–5.3 m^3/s) are plotted in Figure 2; the numbering of the curves corresponds to the treatment trains given in Table III. This analysis includes the energy requirements for unit operation as well as for chemical production, but does not include energy requirements for the transportation of chemicals and the fabrication of the process units.

It is obvious from these figures for the energy requirement that processes needing chemicals have the largest energy consumption, partly because of the transport component included in these values. On the other hand, the energy requirement for biological treatment is reduced as a result of the energy recovered from the digester gas (69,600 kW-h).

Compatibility

It should be kept in mind that the operation of the different components of any treatment system is interdependent and interrelated. Some sacrifice has to be made in the control and performance efficiency when various unit processes are combined together; this is particularly true when physical, chemical and biological processes are integrated together.

Flow Equalization

An equalization basin can be employed to provide both uniform flow rate and more uniform loading rate on the subsequent treatment units. This control of flow rate and concentration assures more reliable and uniform performance of the treatment units. Aeration is provided in equalization basins with a holding capacity of up to 8 h at the design flow.

Flow equalization will have a significant influence on the stability of biological-treatment processes and the uniformity of effluent quality. Use of high-purity oxygen has also improved the performance and stability of biological-treatment units.

Removal of Decomposable Organic Matter

With the use of a mixed bacterial culture, organic materials are converted to oxidized end products (CO_2 and H_2O) and new bacterial cells. Thus, there is an overall reduction in the quantity of organic matter. Biological-treatment plants have been operated very successfully under stable conditions. The microorganisms have an amazing ability to accommodate to variations in hydraulic and BOD loadings and even to adapt to mild toxic substances. However, once the system has been upset it takes a long period to recover. The biological system is also dependent on temperature, pH, dissolved oxygen and alkalinity.

A correctly designed and operated biological-treatment plant for domestic wastewater should produce an effluent with BOD and SS concentrations less than 20 mg/l regardless of the influent concentration. The substances not removed effectively by biological treatment are phosphates, nitrogen, nondegradable organics causing taste and odor, bacteria and viruses, soluble minerals and turbidity. Additional treatment units have to be added if these parameters are to be reduced below the levels normally obtained in conventional treatment plants (Table III).

The effluent COD is composed of bioresistant materials present in wastewater, refractory metabolic by-products and residual compounds. Eckenfelder[8] has shown that the nature and concentration of refractory organics in the effluent after biological treatment bears a relationship to the sludge age or the ratio of food to microorganisms (active) maintained in the aeration basin. An increase in sludge age results in increased residual nondegradable COD. Also, at higher sludge ages, autooxidation of the cell mass will release nitrogen and phosphorus back to solution.

If the biological nitrification–denitrification process is to be used for nitrogen removal, certain modifications will be required in the design of conventional biological-treatment units.

Nutrient Removal

Phosphorus

The most effective method for phosphorus removal is chemical precipitation with ferric salts, aluminum salts or lime, which may be carried out at conventional treatment plants according to one of three alternatives:

1) at the primary treatment stage;
2) in the aeration tank;
3) at the end of the aeration tank.

In the first case, the addition of lime can raise the pH to nearly 10.0, but the natural buffering capacity in aeration tanks due to the production of carbon dioxide lowers the pH to above 7.0. With chemical precipitation at the primary-treatment stage, it is necessary to provide coagulation and flocculation units to achieve maximum efficiency. As shown in Table III, there is a considerable increase in BOD and SS removal.

In chemical precipitation, a considerable amount of sludge is produced. When precipitation is used to improve the removal of organic solids and phosphates in primary or secondary clarifiers, the sludge digestion will be affected by cations and the metal toxicity. Lower suspended solids in the final effluent can be achieved by providing a chemical precipitation unit downstream of the biological process. If chemical precipitation is provided as a separate process, biological and chemical sludges can be separated for the recovery of coagulants; this will decrease the quantity of sludge for disposal.

Nitrogen

In domestic wastewaters, most of the nitrogen is present either as ammonia or in organic form, which gets converted to ammonia during biological degradation. When the carbon concentration in wastewater becomes low and the nitrifying bacteria are populous, this ammonia can be oxidized by bacteria to nitrite and nitrates in the presence of dissolved oxygen.

Conventional wastewater treatment removes only 5–20% of the nitrogen. More effective removal may be achieved by adding one or more of the following unit operations:

a) biological nitrification–denitrification;
b) ammonia stripping;
c) selective ion-exchange;
d) breakpoint chlorination.

Other methods such as land application, demineralization, and algae harvesting, are beyond the scope of this article. Of the four viable systems listed above, only biological nitrification–denitrification and ammonia stripping are considered in this discussion. In certain circumstances, the ion-exchange method may be substituted for ammonia stripping, with similar performance.

The success of nitrification–denitrification depends on a series of biological-treatment units. The biological treatment is sensitive to pH, temperature and toxic material. The units should be designed for the coldest operating conditions and optimum pH (about 8.5 for nitrifiers), and some residual N (up to 2 mg/l) should still be expected in the effluent.[9]

When lime is added for the chemical precipitation of phosphorus, the pH is raised considerably; this raised pH helps in ammonia stripping. However, since phosphorus is necessary for biological treatment, especially nitrification–denitrification, it may become necessary to reverse the order of the processes.

The stripping process can be employed either before or after secondary treatment, to remove the high levels of nitrogen present as ammonia. This process is considered to be the cheapest, simplest and easiest to control among physical–chemical processes,[9] but it has certain limitations:

1) it cannot be operated at ambient temperatures below 0°C;
2) calcium carbonate scales are deposited;
3) pH adjustment between 9 and 11 is necessary;
4) it leads to air pollution (ammonia is released into the atmosphere);
5) nitrification should not be allowed in biological treatment.

Residue

After reduction in the solid content by biological treatment and chemical precipitation, there are still certain inorganic and organic residues remaining in the effluent which may have to be removed. The dissolved minerals are normally of no concern, unless treated effluent is recycled for drinking purposes, and their removal is not considered in this discussion. The suspended inorganic residue in effluent causes turbidity, and the organic residue may be of major concern because it imparts taste, odor and color, and may have physiological effects on the human body.

Turbidity

Filtration through a granular bed is a popular choice for the removal of turbidity. The removal of suspended particles is primarily by a straining action, but the attachment and adsorption of particles on a filtering medium is also possible. The growth which forms on the filter may result in odors if decomposable organic matter is present in the liquid input to the filter beds and in denitrification if nitrates are present.

Tertiary treatment by filtration is strongly affected by the floc overflowing the clarifiers. Surface clogging and cracking may be caused as a result of the rapid accumulation on the top surface of solids produced in the upstream biological and chemical processes.

Organic Residue

Activated carbon removes organic and some inorganic materials from the effluent by the process of sorption. If granular carbon is used, it will also act as a filtering medium. The material removed on carbon surfaces by adsorption is burnt during regeneration and thus causes no sludge-disposal problem.

The generation of hydrogen sulfide in a carbon column due to the bacterial decomposition of sulfates under anaerobic conditions is a serious problem. Efficient biological treatment and filtration of the wastewater prior to carbon treatment can reduce this hydrogen-sulfide generation and also improve the flowthrough rate. The application of chlorine or ozone to the influent and the introduction of air, oxygen or sodium nitrate in the column have been recommended.[9] There is a possibility of some off-gas odors during regeneration.

If powdered activated carbon is used to prevent odor problems it may be applied directly to the mixed liquor in the aeration basin of the activated-sludge plant or used on the secondary effluent. However, this should be followed by filtration.

Chlorination/ozonation

The purposes of chlorination or ozonation in wastewater treatment are disinfection, BOD reduction by chemical oxidation, the elimination or reduction of color and odor and the removal of residual ammonia by chlorination. Both chlorine and ozone have been widely and successfully used, but it should be pointed out that these chemicals are relatively expensive. In recent years, ozone has received more favorable acceptance because of chlorine's reaction with refractory organics to form harmful chloroorganic compounds. It should be remembered that nitrogenous matter in the wastewater is not significantly affected by ozonation.

Breakpoint chlorination may be adopted as the final treatment for very effective disinfection and for the removal of residual ammonia in the event of a sudden drop or failure in the nitrogen-removal operation. This process is not adversely affected by cold weather, but depends on the pH. However, it is a complex and expensive operation, and may add a significant amount of total dissolved solids.

Sludge Disposal on Land

Land disposal systems for sludge may involve its application on soils used for crops and other vegetation, or stockpiling on land. The digested sludge can be applied beneficially on land, improving the physical character and nutrient level of the soil.[10] The land application of organic sludges is a common practice; but chemical sludges can cause some problems, particularly if heavy metals are precipitated and collected in the sludge.

Alum addition may increase the quantity of sludge production by as much as 100%. On the other hand, the solid content of the vaccum filter cake after alum addition is decreased significantly. Thus there can be a substantial increase in the sludge volume for handling.[10]

REFERENCES

1. J. K. Bewtra, "Biological treatment of wastewater," in *Encyclopedia of Environmental Science and Engineering*, Vol. 1, edited by J. R. Pfafflin and E. N. Ziegler (Gordon and Breach, New York, 1976) pp. 109–126.
2. H. I. Ali and J. K. Bewtra, "Wastewater: Physical and chemical treatment," in *Encylopedia of Environmental Science and Engineering*, Vol. 2, edited by J. R. Pfafflin and E. N. Ziegler (Gordon and Breach, New York, 1976) pp. 962–979.
3. R. E. Topping, "Computer program useful in unit operation selection and cost estimation," *Water Sewage Works*, **123**, No. 11, 96–99 (1976).
4. J. C. Young, J. L. Cleasby, and E. R. Baumann, *J. Env. Eng. Div., Proc. A.S.C.E.*, **104**, No. EE2, 289–303 (1978).
5. R. B. Dean and S. L. Forsythe, "Estimating the reliability of advanced waste treatment," *Water Sewage Works*, **123**, No. 6, 87–89 (1976).
6. R. M. Hagan and E. B. Roberts, "Energy requirements for wastewater treatment," *Water Sewage Works*, **123**, No. 12, 52–56 (1976).
7. G. D. Zarnett, *Energy Requirements for Conventional and Advanced Wastewater Treatment*, Pollution Control Branch Publication No. W47 (Ontario Ministry of the Environment, Toronto, Ontario, 1975).

8. W. W. Eckenfelder, Jr., "Interrelationships between biological treatment and physical–chemical treatment," in *Progress in Water Technology*, Vol. 1, edited by W. W. Eckenfelder and L. K. Cecil (Pergamon, Toronto, 1972) pp. 75–184.
9. G. L. Culp, G. M. Wesner, and R. L. Culp, *1974 Lake Tahoe Advanced Wastewater Treatment Seminar Manual* (Clean Water Consultants, P.O. Box 40, El Dorado Hills, California, 1974) pp. 13.1–13.4.
10. W. J. Weber, Jr., *Physicochemical Processes for Water Quality Control* (Wiley, Toronto, 1972) pp. 582–590.

Jatinder K. Bewtra
University of Windsor

RECENT ADVANCES IN HYDRODYNAMIC TREATMENT FOR A BUOYANT SURFACE DISCHARGE

Introduction

In industrially developed countries, more than 60 % of industrial freshwater is utilized as cooling water. Sea water is also used, especially in the coastal industrial zone. Almost all of the industrial cooling water is accounted for by cooling water in steam-based power plants.

The total world energy consumption in 1970 was approximately 7 billion tons, as converted to the equivalent weight of coal, and is apparently increasing at the rate of 4–5 % per year. About 30 % of this energy is in the form of electric energy, generated from fossil fuel, water power and atomic energy. The recent predictions that fossil fuel and water power cannot be expected to make as large a contribution in the future mean that the dependence on atomic power as a source of electrical energy will steadily increase. According to a prediction by the International Atomic Energy Agency, the electrical-energy output from atomic power plants will rise from its 1970 level of 1 million MW to reach 3 million MW in 1985 and 7 million MW in 2000 and, accordingly, the share of atomic power in the total electrical-energy production will increase from 2 % to 27 and 57 %, respectively.

A cooling system is indispensible for steam-based power plants. In the case of a once-through cooling system, a 1000-MW output requires 35–40 m^3/s of cooling water for a plant utilizing fossil fuel, and 60–65 m^3/s for an atomic power plant. In a steam-based power plant, the turbine is driven by the energy of steam, which derives from heat energy generated in the boiler or nuclear reactor. After passing through the turbine, the steam is conducted to a condenser, where it is cooled to form water once again. This process is necessary in present power-generation systems; unavoidably, therefore, a large amount of heated wastewater is produced in the cooling system. The accompanying energy loss by heat transfer is as much as 44–47 % for fossil fuel and 62–66 % for nuclear fuel.

Whatever cooling system is adopted, the heat in the thermal wastewater will ultimately radiate into the atmosphere. The solar energy which the Earth receives is estimated at 1800 billion MW, and the total heat energy radiated from steam-based power plants, estimated at 10^{-5} of the solar total, is quite small from a global viewpoint. Hence, if the thermal radiation from the Earth's surface is *adequately* distributed in space, the thermal effect on the environment might be negligible. However, if the temperature of the effluent water is quite high, and diffusion takes place in a limited area, it can be easily supposed that the local changes in the environment will seriously affect the ecological system.

In evaluating the effects on the environment, two approaches are adopted. One is to study the hydraulic and thermodynamic characteristics of the thermal effluent, the results being used to establish an analytical model for the prediction of the current and temperature distributions in the water concerned. The other is to study the biological and ecological characteristics of the water, in order to estimate the effect on aquatic plants and animals of short-term changes in the water temperature or other physical factors and long-term variation in the water quality. Such investigations are now being conducted, so that the normal environment of aquatic biota can be maintained, but as yet no more than fragmentary information is available regarding the effect of heat on the ecology of the hydrospheric environment. This problem will only be resolved on the basis of an integrated approach involving the accumulation of field observations, simulation analysis and theoretical analysis.

The present review deals mainly with recent developments of theoretical analysis and with fundamental hydraulic experiments on surface discharge. In addition, some new approaches to turbulence analysis are described; these give new insights into the actual conditions of microscopic turbulence in the diffusion of a surface discharge.

1 Practical Notes on Cooling Systems

1.1 Intake of Cooling Water

In countries with modern power plants, the amount of cooling water used tends to be around 10^5–10^6 m^3/day; in the United Kingdom and Japan, the corresponding figure is 8×10^7 m^3/day, and in the United States it is 44×10^7 m^3/day. The location of the plant determines whether the water required is brought from rivers, lakes or oceans.

In the United States, many steam-based power plants have been developed inland, and obtain their cooling water from large lakes and rivers. To ameliorate the thermal effects of these systems on aquatic life, the effluent temperature is regulated, and the use of cooling-tower systems has been actively promoted. In parallel with those efforts, consideration has been given to the construction and planning of coastal power plants, and to the adoption of a closed-cycle cooling system.

In the United Kingdom, steam-based power plants located inland depend on rivers for their cooling water. Difficulties in obtaining sufficient water have led to efforts to introduce water-recirculation systems, and many cooling towers have been constructed. In the case of coastal power plants, a once-through system is usually adopted.

In Europe, where many rivers cross national boundaries, the establishment of effluent temperature standards has been the subject of considerable discussion and, in general, power plants now under construction or at the design stage have provision for the inclusion of cooling towers.

In Japan, because of difficulties in obtaining an adequate supply of river water, all steam-based power plants are located on the coast or in industrial districts near river mouths, and adopt a once-through system.

1.2 Types of Cooling System and Their Operation

Once-through cooling system In this system, the cooling water is drawn from a body of water and, after passing through the condensers, is discharged back into the same body of water. This system is adopted at inland power stations when adequate amounts of water may be obtained from rivers or lakes and, of course, at coastal power stations, where the sea provides an ample source of water. To avoid physical interaction between the intake point and the outflow, they are generally located as far apart as possible. The discharge system must be designed so as to produce thermal equilibrium as rapidly as possible in the limited water area. The simplest such system is a surface jet discharge; where geographical and meteorological conditions permit, submerged jets, submerged multicomponent diffusers and other similar devices may also be used to good effect.

Closed-cycle cooling system Where there is difficulty in obtaining sufficient water, and in maintaining the normal environment in the water body concerned, the cooling-tower system is often adopted. In this system there is no thermal effect on the water environment, because the heat is emitted directly into the atmosphere after the heat transfer in the cooling tower. However, the construction costs are, in general, high, and so, to achieve the most efficient disposal of waste heat, cooling towers are often used in combination with other cooling systems.

One such is the cooling-pond system, with recirculation of the water. Recirculation is also involved in the case where sea water is drawn from a semiclosed body of water. Surface discharge to a cooling pond is a common disposal method, but careful structural design is necessary to avoid problems with the heat emission from the limited surface area and the intake of cooler water.

1.3 The Mixing of Surface-discharged Water

As noted above, surface discharge is employed in once-through cooling systems and in cooling-pond systems. The physical effects of surface discharge on the mixing, dilution and heat emission in the water body have been extensively investigated. The temperature of the discharged water is 5–7°C higher than the ambient temperature and so, being ligher, this water spreads thinly and widely in a shallow layer of the water body. The diffusion and cooling of the water are governed by three fundamental processes: heat transport by the effluent flux itself; heat dilution by mixing with the ambient water; and heat emission into the atmosphere.

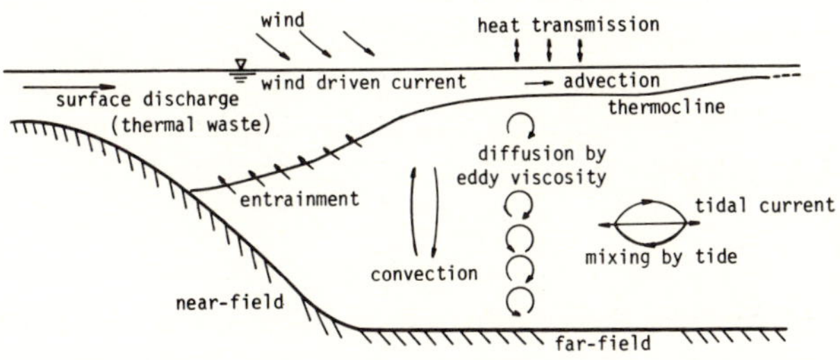

Figure 1 Mixing processes associated with a surface discharge.[24]

The mixing processes associated with diffusion of the discharged water include such fluid-dynamic phenomena as entrainment from the lower layers of the water body, diffusion by turbulent eddy viscosity and, in the case of a coastal plant, tidal motion. These processes are summarized in Figure 1. Thus the mixing mechanism is complex, especially for coastal power plants, where tidal motion is likely to be dominant not only in mixing but also in the formation of the region of thermal influence in the sea. In the future, moreover, the number of coastal plants must be expected to increase.

2 Buoyant Jets

Consider a flow released into a still or moving water body. In general, such a flow is defined as a jet, and its motion is characterized, basically, by the mass and momentum flux. A jet is unstable and sensitive to the external circumstances; nevertheless, its motion spreads, preserving its own mixing zone. When the difference in density between a jet and a surrounding water body is significant, the effluent jet is also susceptible to gravitational influences: it is then defined as a *buoyant jet* or a *forced plume*. The most important dimensionless factors governing the velocity and density distributions in the plume and the stability on the boundary are the internal Froude number (or densimetric Froude number) and the Reynolds number:

$$F_0 = \frac{U_0}{\sqrt{[(\Delta\rho/\rho_w)gD_0]}}, \qquad \mathrm{Re}_0 = \frac{U_0 D_0}{\nu_0}, \tag{1}$$

where U_0 is the velocity at the outlet; D_0 is a characteristic length corresponding to the shape of the outlet; $\Delta\rho = \rho_w - \rho_0$; ρ_w is the density of surrounding water; ρ_0 is the initial density of the plume; ν_0 is the coefficient of kinematic viscosity of the plume. When $F_0 \to \infty$ the plume turns into a jet with dominant momentum flux, and when $F_0 \to 0$ it becomes a pure plume governed by the buoyancy flux alone. Another dimensionless factor, the Richardson number

$$\mathrm{Ri} = \frac{1}{F^2}, \tag{2}$$

is available to describe the stability of stratification, which is given by the ratio of the buoyancy and the inertia to the intensity of turbulence. On the other hand, if $\mathrm{Re}_0 > 10^4$, Ri may be neglected in discussing the characteristics of the time-averaged velocity and density from the viewpoint of experimental results.

A plume has unique characteristics with respect to mixing and diffusion; the following factors will be considered:

1) the geometric shapes and scales of the outlet;
2) the densimetric properties of the plume;
3) the magnitude and direction of action of the momentum and buoyancy flux;
4) topographic factors such as the shore line, depth, etc.;
5) limnological or oceanographical and meteorological phenomena such as the current, tidal motion, wave motion, stratification, etc.

Tollmien identified three types of free turbulence for a jet: a free jet boundary, a plane jet and a round jet[1]; since then, considerable research has been done, by, among others, Görtler[2] and Albertson *et al.*[3] The method of analysis adopted was to use the mixing-length coefficient or the exchange coefficient in the governing equations, in order to derive unknown quantities of the jet. The results obtained were in agreement with experimental data, and these theories could serve to progress the problems to new levels. On the other hand, the research on stratified-density current may be divided into two groups: (1) work dealing with the amplification or attenuation of internal waves, including a theoretical discussion of the idealized velocity and density distributions for both viscid and inviscid fluids[4–7]; (2) studies dealing with the entrainment of ambient water through a stratified boundary, with the experimental determination of the entrainment constant, the value of which seems to be proportional to the relative velocity in two stratified layers.[8,9]

Both results suggested that the local Richardson number might be an important parameter of the entrainment mechanism.

The use of turbulent eddy viscosity and entrainment constants provides a convenient explanation of the mixing volume from the ambient water body to the jet zone without the need to know how the individual turbulent eddy moves and contributes to the mixing process. With the aid of the conservation laws for mass, momentum and dissolved material, this approach has led to the successful solution of several practical problems. One example is the theoretical analysis of a plume. There have been several theoretical or experimental studies of horizontal and vertical plumes released into a uniform water body or a stratified reservoir.[10–13] The entrainment constant, which depends on the turbulent characteristics and on the local Richardson number, is treated as an empirical fixed value in the range 0.036–0.082 in these studies, but its explicit value is unknown.

In studies of plume diffusion in a moving water body,[14–17] the plume path, its width and the dilution ratio are obtained analytically by introducing two characteristic lengths that are simply related to the momentum at the outlet and the buoyancy flux. The dilution ratio is generally proportional to the path length and the entrainment constant varies from 0.17 to 0.53, which is one order of magnitude greater than that in still water. Thus, the sensitivity of the dilution process to the ambient flow states is evident.

3 The Diffusion of Surface Discharge

3.1 General Remarks

It is convenient to consider the cooling of a surface discharge separately in two distinct regions: the near field and the far field.[18] In the near field, which is called the *initial mixing region*, the momentum flux of the effluent water body forms itself into a prevailing jet flow from the outlet. The total sectional flow rate rapidly increases downstream in this field because of dominant entrainment from both lateral and lower water bodies with irregular eddy motion, and there is a rapid decrease in the difference in velocity and density between the jet flow and the ambient water. However, as the jet flow has too much momentum, the motion of the flow is conservative, in general, though it receives intense entrainment from the ambient water body; thus, a jet is said to be a self-preserved flow accompanied by entrainment. In the far field, called the *off-shore diffusion region*, the initial momentum of the jet is dispersed in space and hence the discharged water body gradually loses the characteristics of a jet and becomes a mild density current. In that process, the effluent water spreads thinly over a tongue-like region, in such a manner that it develops the lateral mixing due to the horizontal turbulent eddies in the water region, while vertical mixing occurs according to the balance condition between the stratification, which tends to inhibit the mixing, and the instability, which tends to force the mixing.

3.2 Mathematical Modeling of Surface Buoyant Jet inside the Near-field Region

The fundamental equations governing the diffusion of a buoyant jet are the conservation laws for the mass, momentum and buoyancy. Basically, the theoretical method combines the analysis of a turbulent jet in a uniform density fluid with an analysis of the buoyant effects due to the difference in density. The analysis in the established turbulent region in three-dimensional space rests on the following assumptions:

1) that the behavior of the jet is symmetric about the jet axis, and its time-averaged velocity and buoyancy distributions are similar in shape on an arbitrary section of the jet.

2) that the mean velocity is independent of time and the flow is fully turbulent;

3) that the Boussinesq approximation is valid, i.e., the density variation affects only the gravity term;

4) that the water pressure is hydrostatic;

5) that the receiving water region extends infinitely into the offshore region.

Considering a surface jet (Figure 2) of density ρ_a, discharged from an outlet ($b_0 \times h_0$) with velocity U_0 and density ρ_0 ($\rho_a > \rho_0$), assumption (1) means

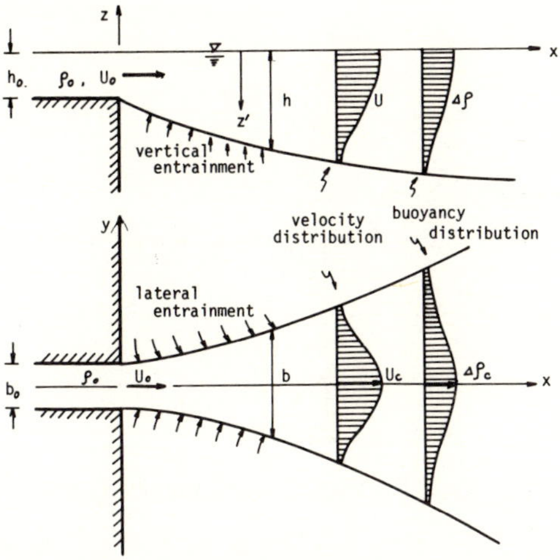

Figure 2 Buoyant surface discharge—near-field region.

that the velocity and density distributions may be described by similar curves, which are empirically given by Gaussian curves:

$$\left.\begin{aligned} U/U_c &= \exp(-\eta^2)\exp(-\zeta^2) \\ \Delta\rho/\Delta\rho_c &= \exp(-\eta^2/2)\exp(-\zeta^2/2) \end{aligned}\right\} \tag{3}$$

where U and $\Delta\rho = g(\rho_a - \rho)/\rho_a$ are the velocity and buoyancy at an arbitrary point of the jet; U_c and $\Delta\rho_c$ are the velocity and buoyancy on the jet axis at the water surface; $\eta = y/b_{1/2}$; $\zeta = z/h_{1/2}$; $b_{1/2}$ and $h_{1/2}$ are the characteristic width and depth, respectively, which give, in general, the width and depth at the points where $U = U_c/2$.

Integrating the equations of volume continuity, motion and buoyancy from 0 to ∞ with respect to z, and from $-\infty$ to ∞ with respect to y, three conservation equations are obtained, as follows

$$\frac{dQ}{dx} = \int_0^{\zeta} V_e \, d\zeta + \int_{-\eta}^{\eta} W_e \, d\eta, \tag{4}$$

$$\frac{dM}{dx} = \rho_a \frac{d}{dx} \int_A \left\{ \int_0^{\zeta} \rho \, d\zeta \right\} d\zeta \, d\eta, \tag{5}$$

$$\frac{dB}{dx} = 0, \tag{6}$$

where V_e and W_e are the horizontal and vertical entrainment velocities, respectively, and the physical quantities to be conserved are defined as follows: volume flux

$$Q = \int_A U \, d\eta \, d\zeta; \tag{7}$$

momentum flux

$$M = \int_A \rho_a U^2 \, d\eta \, d\zeta; \tag{8}$$

buoyancy flux

$$B = \int_A \Delta\rho U \, d\eta \, d\zeta. \tag{9}$$

Following Morton,[8] the entrainment coefficients α_y and α_z are defined by the relations

$$\begin{aligned} V_e &= \alpha_y U_c, \\ W_e &= \alpha_z U_c, \end{aligned} \tag{10}$$

which state that V_e and W_e are proportional to the surface velocity on the jet axis. The coefficient α_y is expected to be proportional to the spreading rate of the width dB/dx from the viewpoint of mass continuity. In addition, the entrainment volume from the lower part of the water body decreases as the vertical stratification becomes settled, according to experimental results[9]; these indicate that the Richardson number,

$$\mathrm{Ri} = \Delta\rho_c \frac{h}{V_c^2} = g \frac{\rho_a - \rho}{\rho_a} \frac{h}{U_c^2}, \tag{11}$$

plays a dominant role in the entrainment, and the relation between α_z and Ri is as follows

$$\alpha_z = 0.075\left(\frac{2}{1 + \mathrm{Ri}/0.85} - 1\right)^{1.75}. \tag{12}$$

Hence, at the critical Richardson number $\mathrm{Ri}_{cr} = 0.85$, $\alpha_z = 0$. As the surface discharge spreads offshore, Ri approaches Ri_{cr}, so that there could exist a limit $x = X_{\mathrm{Ri}}$ which separates two different mixing regions.[34]

Since Eqs. (4)–(6) are not enough to determine the four unknown factors

$$U_c(x), \quad \Delta\rho_c(x), \quad b_{1/2}(x), \quad h_{1/2}(x),$$

it is necessary, as an analytical device, to introduce a new characteristic relation associated with the equation of motion in the y direction: this may be done by assuming a distribution profile for the velocity V in the y direction,[19] by determining this profile from the equation of continuity[20] or by expressing the spreading width as a function of the Richardson number.[21,22] Combining the results of all these approaches, the following equation is obtained

$$\frac{db}{dx} = \varepsilon + \left(\frac{db}{dx}\right)_B, \tag{13}$$

where ε is the rate of lateral spreading for a jet in uniform-density water, approximately equal to 0.22[23]; $(db/dx)_B$ corresponds to the gravitational stability of the density difference for a surface discharge. In all the above approaches, the similarity function, the entrainment constant and the rate of lateral spreading are established on the basis of experimental results, so that the agreement between the theoretical and experimental results must be verified.

3.3 Mathematical Modeling of the Diffusion of a Surface Discharge in the Far-field Region

At large distances from the outlet ($x \gg x_{Ri}$), the turbulence of the jet is not dominant over the buoyancy, and the entrainment will be lost because of the establishment of a stable thermocline. Moreover, because of lateral spreading, the thickness of the surface layer and the velocity in it are small. In this case, the mass and the contaminants, including heat, in the surface layer are carried downstream by its own flow; this is called *advection* or *horizontal convection.* At the same time, they are mixed or diluted with the lateral ambient water as a result of the eddy diffusivity of the lake or tidal current, although vertical entrainment decreases markedly. On the other hand, thermal diffusion continues unabated, chiefly in the form of heat emission from the water surface.

Field observations of the surface discharge of thermal waste in the sea have yielded several interesting results, as follows.[24]

1) The discontinuous plane of the density forms a horizontal stratified layer of thickness 2–3 m. This state is much the same in any sea region for any of the latest power plants.

2) This thermocline is stable even if the tidal level varies with time; moreover, it is not broken down by wind-driven current with a wind velocity of 5–10 m/s, though its level sinks down slowly as a result of the mixing in the surface layer.

3) When an evident thermocline is formed in the water body, the thermal wastewater does not mix with the bottom water through the thermocline but tends to mix with the surface water. Therefore, the horizontal thermal

diffusivity has been found to be at least 50 times larger than the vertical value.

4) According to an empirical relation between the flow rate of the effluent and the area that is influenced by its thermal wastewater, an area of approximately 1 km^2 is 1°C higher than the ambient water in the case of an effluent flow rate of 50 m^3/s and a temperature difference of 5°C between the effluent and the ambient water; an interval of more than 50 hours is required to reach such a state.

On the hypothesis that the stable stratified layer might exist for a long period of time and over a broad area, Wada has proposed an analytical method for the two-layer model.[24] The fundamental equations are obtained by integrating the equations of motion, continuity and diffusion in the surface layer from the internal boundary to the water surface—and in the bottom layer—from the bottom to the internal boundary. Omitting the diffusion equation in the bottom layer because the density is constant, the equation in the surface layer is given by

$$\frac{\partial T}{\partial t} + U_j \frac{\partial T}{\partial x_j} = \frac{1}{h} \frac{\partial}{\partial x_j} K_j h \frac{\partial T}{\partial x_j} + \frac{Q_0}{c\rho h}, \tag{14}$$

where $j = 1, 2$ for the x and y directions, respectively; U_j are the velocity components in the x and y directions; T is the water temperature in the surface layer; h is the water depth in the surface layer; K_j are the thermal eddy diffusivities; Q_0 is the heat gain or loss for the surface layer of the water basin; c is the specific heat of water. The heat gain or loss Q_0 in Eq. (12) may be expressed in the form

$$Q_0 = Q_s - Q_b + Q_n + Q_e = Q_1 - Q_2 T, \tag{15}$$

where Q_s is the absorption of radiation from the sun and the sky; Q_b is the back radiation from the water surface; Q_n is the convection of sensible heat from the atmosphere; Q_e is the evaporation. As an illustration, numerical solutions for the fundamental equations, including Eq. (12), are given in Figure 3, which shows isotherms of the surface discharge for different diffusivities in a water field with an alternating current as a result of the tide parallel to the coastline; Figures 3a and 3b show the cases of ebb tide and flood tide, respectively. In this computation, two models of the diffusion are used: the continuous curves are for the case where the diffusivity K_x is constant offshore, while the dashed curves are for the case in which K_x increases linearly offshore. It is noteworthy that the dashed curves follow the tidal cycle more sensitively, which means that, depending on the choice of diffusivity, the principal difference between the two models is found in the water-temperature distribution. On the basis of these results, Wada suggested that, when a large amount of cooling water is discharged, it is

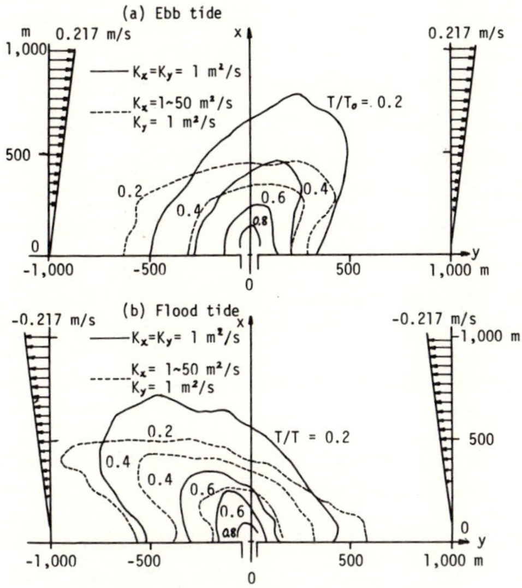

Figure 3 Temperature distribution in a water basin in the presence of a current[24]: (a) ebb tide; (b) flood tide.

necessary to consider a diffusion model in which large-scale eddies contribute to the diffusion process in the far field of the sea. In fact, various scales of eddies and complex unsteady currents appear as the distance from the coastline becomes large. Wada *et al.* have discussed the independent effects of currents and diffusion on the overall diffusion process, considering the time scale and the spatial scale obtained from the statistical data of field observations.[25] In dealing with the short-term variation of the diffusion, the tidal current may be regarded as a mere stream, but for the long-term variation the current must be regarded as large-scale eddies in terms of its contribution to the diffusion. It is important, therefore, to have a correct understanding of the space and time scales involved and to make an appropriate choice of model in the simulation analysis.

4 Experimental Research on Surface Buoyant Jets

4.1 General Remarks

In spite of many tests of the analytical model of a surface buoyant jet on the basis of laboratory experiments and field observations, the hydraulic properties of entrainment remain unclear. Many experiments have been carried out to gain an understanding of entrainment and to verify analytical work, although the detailed aims of the experiments have varied.

In the early stages, the emphasis in the experiments was on the investigation of buoyant effects in a horizontal buoyant jet, and the characteristics of the velocity and density distributions became clear to a certain extent. Densimetric Froude numbers and Reynolds numbers of 1.4–16.1 and 3100–21,300, respectively, were used in these experiments.[3,19,26–32] The experiments of Tamai;[33] Koh[34] and Stefan[35] on a two-dimensional horizontal buoyant jet deserve special mention. All the experiments have relevance to the reconfirmation of the assumptions in Section 2, the evaluation of the similarity function and the application of the entrainment constant to three-dimensional problems. The results may be summarized as follows.

1) The velocity and density distributions may be approximated by a Gaussian curve.

2) The horizontal and vertical spreading, which appear as an increase in the standard deviation of the Gaussian distribution, are different.

3) It is difficult to obtain an exact prediction of the width of the lateral spreading on the basis of the entrainment constant, but it is possible to predict the mean values of hydraulic quantities along the jet axis on this basis.

It is clear from these studies that the entrainment constant plays an essential role in the analysis of a horizontal buoyant jet. At present, however, only the relation between the entrainment constant and the overall Richardson number is known experimentally, and there is little information on the densimetric Froude number or the Reynolds number at the outlet. This is partly because of difficulties of experimental technique: continuous temperature measurement is not easy because of the heat emission from the water surface; the response characteristic of the instrument measuring the saline concentration is not very good; boundary effects on the measuring data in the experimental basin are involved; and so on. In addition, there are only a few experiments on the simultaneous measurement of turbulence of the velocity and concentration,[36,37] even though this is important for an understanding of the essential relation between the turbulence mechanism and entrainment. The next stage in the research must be to obtain a better understanding of the hydraulics of entrainment itself, in order to devise an engineering solution that allows the temperature difference and the influence zone of the discharge to be reduced.

4.2 Effects of the Reynolds Number

From an empirical viewpoint, it may be stated that the velocity profile in a buoyant jet is not substantially affected by the Reynolds number as long as the flow is turbulent. Nevertheless, it remains to ask whether the Reynolds number plays a significant role in the entrainment process in spite of turbulent phenomena. In observations of float trails on the water surface,[37] it was found that the float patterns may be grouped in three different categories according to the magnitude of the Reynolds number. Type 1 is the irrotational flow pattern, in which floats circulate slowly along the boundary in the opposite direction to the jet flow and are finally engulfed by the jet. In this case, $Re_0 \simeq 2800$. With increase in Reynolds number, this pattern is replaced by the transitional type 2, in which floats approach close to the jet axis and move downstream, rotating slightly near the jet boundary, and are then engulfed into the jet. Lastly, in type 3, with $Re_0 > 11{,}000$, floats also approach close to the jet axis and are entrained as they are whirling.

Thus it seems that the Reynolds number is directly related to the development of the mixing layer or the eddy motion in that layer, and plays and important role in characterizing the entrainment process, although it has little effect on the mean values of hydraulic quantities inside the buoyant jet.

4.3 Effects of the Densimetric Froude Number

A plot of the lateral spreading width is shown in Figure 4, where the width is measured from a point at which the approaching velocity would be a minimum near the nominal jet boundary. The plotted points are considerably scattered, but it is clear that the lateral spreading is influenced by the buoyancy, i.e., the gradient of the width db/dx lies in the range 0.3–0.8 in

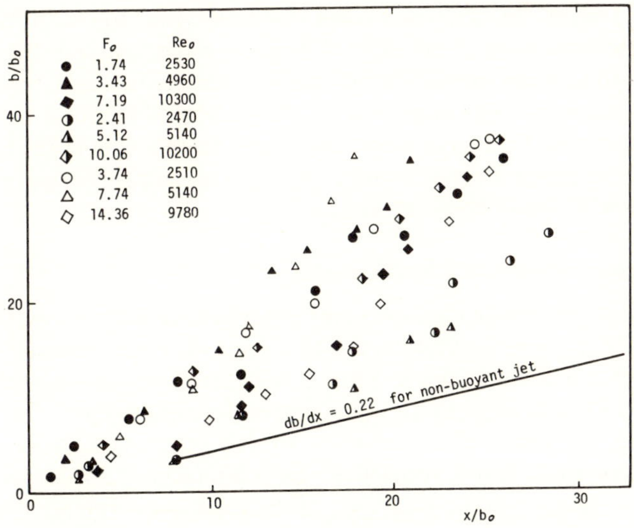

Figure 4 Width of surface buoyant jet for various F_0 and Re_0.

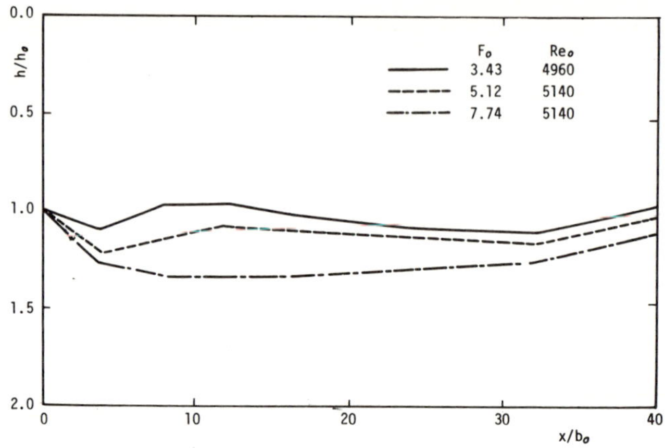

Figure 5 Depth along buoyant jet axis.

the case of the surface buoyant jet. This is rather larger than the value for a nonbuoyant jet, $db/dx = 0.22$.[23] In addition, the gradient seems to be large when the densimetric Froude number is high.

On the other hand, Figure 5 shows the distribution of the surface-layer depth h, the bottom of the layer being defined at a point where there is a sharp change in the local Richardson number Ri calculated from the time-averaged values of the velocity and density. It is interesting to note that the variation of the depth dh/dx is extremely small, and no vertical spread is seen in the region $x/b_0 > 16$. The same tendency is observed in the velocity profile $U(z)$. Thus, the buoyant surface discharge tends to retain its gravitational stability because of the slight difference in density, and this restrains the surface flow from spreading vertically while at the same time forcing the lateral extension. The magnitude of the density difference is represented by the densimetric Froude number, and in this sense it is possible to suggest that the densimetric Froude number may be an important parameter in connection with the vertical entrainment.

4.4 Turbulence-intensity Characteristics

Since the first systematic measurements of turbulence in a nonbuoyant jet,[38] many such data have accumulated. There have also been measurements of the turbulence characteristics as a function of the Reynolds number and the densimetric Froude number for a buoyant surface discharge.[36,37]

Denoting the mean velocity and the velocity fluctuation in the direction of the jet axis by U and u, respectively, it is interesting to examine the distribution of the turbulence intensity $\sqrt{\overline{u^2}}/U$ (Figure 6). The intensity at the outlet is less than 10% but it increases markedly, reaching 40% at a relative

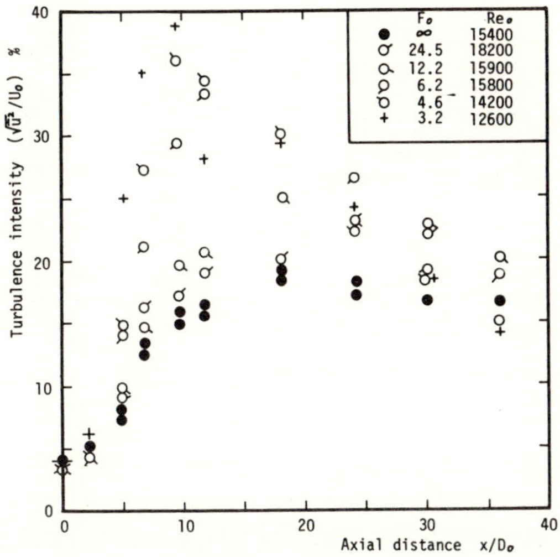

Figure 6 Turbulence intensity along jet axis in surface buoyant jet.[36]

distance from the outlet of $x/h_0 = 10$. As the decay of the mean velocity along the axis is not very great, this marked increase is worthy of note. The tendency is more pronounced for a jet with a low value of the Reynolds number, and the peak value of the intensity is closer to the outlet. This is consistent with the observation that, for low values of the Reynolds number, the potential core of the jet is shortened and accompanied by brisker mixing.[39]

Turning now to the effect of the densimetric Froude number, F_0 is larger in the range $x/b_0 < 10$. It is known that, in the case where F_0 is small, the mean velocity along the jet axis decays rapidly, because the stratification becomes more stable and the lateral spreading becomes more active. However, even if this is taken into account, the increase in turbulence intensity seems to be too large. It seems that this is associated with the strengthening of turbulence due to the extra coupling of the increased lateral entrainment. From the above it may be concluded that, although the mean profiles of the jet are self-preserving and are geometrically expressed as similar Gaussian curves, the characteristics of the fluctuating velocity are significantly influenced by both the Froude and Reynolds numbers.

Figure 7 shows the vertical variation of the velocity fluctuation u and the salt-concentration fluctuation c (mean salt concentration C) measured simultaneously at $x/b_0 = 16$ for a jet with $F = 6.4876$. The local Richardson number is evaluated numerically from the formula

$$\mathrm{Ri} = -(g\mathrm{d}C/\mathrm{d}z)/[C(\mathrm{d}U/\mathrm{d}z)^2],$$

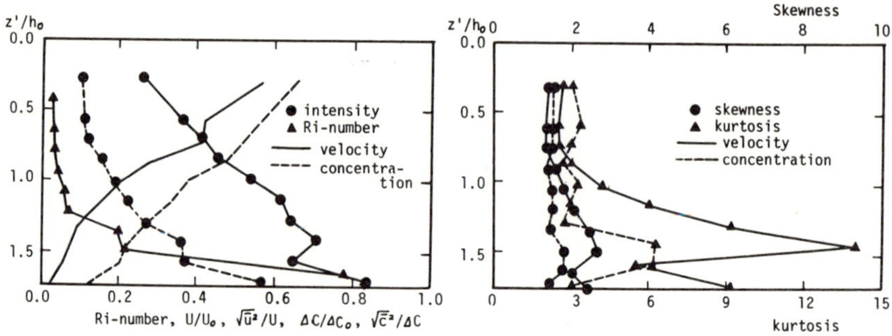

Figure 7 Vertical distribution of mean and fluctuating values of velocity and concentration measured at $x/b_0 = 16$, $y/b_0 = 0$.

which indicates the local stability of the interface. The vertical distribution of Ri shows a rapid increase near the depth $z'/h_0 = 1.43$–1.57. Note that this depth corresponds to the point at which the vertical gradient of the mean velocity varies and the relative intensities of u and c increase, too. Hence, the layer at which the rapid increase in Ri occurs may be regarded as an interface lying horizontally in the jet. With respect to the velocity and concentration fluctuations, statistical properties such as the skewness S and kurtosis K are compared in the same figure in order to gain a quantitive understanding of the intermittent properties near the interface. Although the skewness of both u and c distributes uniformly below 2.0, the kurtosis of u increases abruptly to almost 14.0 at the depth at which the rapid increase in Ri is observed. As the variance of the turbulent intensities is comparatively small in the vertical direction, it will be assumed, as an approximation, that the deviation from the mean distributes in a wide range and the turbulent structure becomes coherent near the interface.

5 Entrainment and Mixing

5.1 *The Mechanism of Mixing*

The mixing of water quality through the interface of a buoyant jet is defined by the entrainment, and several types of entrainment may be distinguished, according to the primary generating factors of the turbulence near the interface, for example: the diffusion type—entrainment with small eddies which occur as a result of frequent large disturbances or instability of the interface; the rolling-up type—entrainment due to the breaking of internal gravity waves; and the convection type—entrainment due to the engulfing of large-scale eddies. For a buoyant surface discharge, the turbulence generated by the imbalance between buoyancy and momentum would be a trigger leading to the mixing of water quality. Entrainment can be regarded as the intake of a water body that a turbulent zone recieves from a nonturbulent zone or, from another point of view, as the intrusion from a nonturbulent zone into a turbulent one. Therefore, the process that is responsible for the intake or the intrusion cannot be adequately explained by means of the concept of turbulence transport, which is constructed from extra momentum and extra buoyancy; it is more reasonable to consider that the turbulent eddies result in some action on the interface, then transform or entangle the interface and consequently produce the mixing of water. In investigating the mixing mechanism, therefore, the first step should be to sample the representative scales of the turbulence, which seems to be intimately related with the mixing motion.

As already noted, there are numerous studies on the stability problem in stratified shear flow, but in many of them the emphasis is on problems of the internal wave or the initial mixing, and quantitative consideration of the entrainment is impossible.

5.2 The Entrainment Constant

The earliest experimental study on entrainment through an interface was by Keulegan.[7] He defined the Keulegan number, which indicates the stability of an internal wave in a viscous fluid, and found that, when that number is less than a critical value, the internal wave breaks down and the lower layer of water begins to entrain into the upper layer. The entraining volume is named the *entraining velocity* because it has the same dimension as velocity, i.e., the volume passing across unit area in unit time. The entraining velocity obtained by Keulegan seems to be approximately proportional to the mean velocity in the upper layer.

Morton *et al.* regarded entrainment as a phenomenon in which the ambient fluid is engulfed into the turbulent zone; this flow is directed perpendicular to the main flow direction and its velocity is proportional to that of the main flow.[8] The constant of proportionality is called the *entrainment constant.* If the flow is in a state of local equilibrium, the entrainment constant is defined by local scales of length and velocity. Subsequently, Ellison and Turner introduced the overall Richardson number as an index of stability, that is, the ratio of the buoyancy to the inertial force.[9] On the basis of experimental data for a horizontal buoyant jet, they established a relation between the entrainment constant E and the overall Richardson number Ri—Eq. (12). Wada and Katano also arranged their experimental data in a similar manner, but argued that there is not so clear a relationship between E and Ri.[40] They pointed out that the slope of the interface $\mathrm{d}h/\mathrm{d}x$

can also be an important factor and concluded that the entrainment would be independent of the Richardson number and that no entrainment is observed if $dh/dx < 0.04$.

On the basis of the theory of wind-wave generation, Phillips proposed a momentum-transfer model in the vertical direction for the shearing flow of the ocean.[41] He argued that the spectrum with respect to the vertical velocity in the stratified flow is dependent on the Brunt–Väisälä frequency, $N = (-g/\rho \partial\rho/\partial z)^{1/2}$, and calculated the entraining volume in the vertical direction using the ratio of the shear stress in stratified and nonstratified flow. The entrainment constant is given by

$$E_s = E_0 \exp(-\beta/F_a^2), \tag{16}$$

where E_0 is the entrainment constant for nonstratified flow and F_a is the local densimetric Froude number. Though the derivation is different, this expression is similar to the Ellison–Turner equation.[9]

5.2 The Structure of Turbulence

Because the water-quality mixing depends on the turbulent characteristics in the neighborhood of the interface, the measurement of turbulent quantities with respect to velocity and physical scalars such as temperature and concentration has been actively pursued along with the development of appropriate instrumentation. From the results, frequency ranges in which the spectral curves are simulated by $f^{-5/3}$, f^{-3} and f^{-5} have been ascertained.[40,42–45] All these measurements have been carried out for turbulence in stratified flow with a mild density gradient. There has been little discussion of the internal turbulent structure and the possible influence of the density gradient or the spectral profile, the upper limiting wave number in the buoyancy subrange, etc. As far as can be determined, only Tamai and Nishimura have discussed the entrainment process in the context of the spectral profile.[46] They defined the wave group which contributes to the mixing as consisting of waves with a wave number greater than the critical wave number k_m for the stability of a stratified flow with a jet-type density profile.[47] In the buoyancy subrange obtained theoretically by Lumley,[43] they integrated the turbulent energy spectrum from k_m to k_b, which is the upper critical wave number available for the buoyant effect, to obtain the total energy with respect to the unstable waves. They found that, in the upper layer, the turbulent energy contributing to the mixing is inversely proportional to the Richardson number; this functional relationship is similar to that obtained experimentally by Ellison and Turner.[9]

The investigations described above, however, are often based on the dimensional analysis of turbulent quantities. It is of great use to discuss the physical meaning of each term in the balance equation with respect to turbulent components, but this is not sufficient to solve the entrainment process. It is important to discuss the relations between the Reynolds number, densimetric Froude number, Reynolds stress and $\overline{\rho w}$, where the last term is the vertical flux of water quality, the first and second are external parameters and the last two are internal microcharacteristics to be measured with turbulence sensors. In particular, the correlation $\overline{\rho w}$ plays a role from the viewpoint of the dynamic equations, generating the density fluctuations by acting on the potential energy. Experiments on the behavior of $\overline{\rho w}$ have shown that, in the region close to the edge of the turbulent zone in a buoyant surface jet, the coherence of the velocity and concentration fluctuations is restricted to a narrow band on the low-frequency side.[48] The cross-correlation among them is ± 1.0, with a delay time. This suggests that the mixing of the water is associated with large-scale eddies, but the discussion of these eddies could not be immediately related to the previous discussion of the dynamic equation.

5.3 Large-Scale Eddies

The recent discovery, in investigating turbulent shear flow, that turbulence has a coherent structure has brought about a 180° turn in the direction of research. In jet or open-channel flow, the strucure of turbulence consists not only of irregular or random turbulence but also of the large-scale eddies called quasiordered turbulence. The behavior of such eddies, the scale and period of which seem to be equivalent statistically to those of the time-averaged quantities for the stream, was minutely investigated by Lauter.[49] The authors have also observed them in an experiment by a visualization technique on a two-dimensional horizontal buoyant discharge. In the neighborhood of the interface, eddy motion, the scale of which corresponds to the thickness of the mixing layer, is seen to generate, grow and entrain the ambient water, and repeated pairing and bursting of eddies is observed. On the record of velocity over time at that location, the intense turbulence peculiar to free turbulence and the loose fluctuation appear alternately. This alternation, known as *intermittency*, is quite different from the behavior in the turbulent zone of the jet. On the hypothesis that the entrainment is excited by this intermittent variation of short duration, the discovery of large-scale eddies would strongly suggest that if the coherent turbulence structure were to be established by the stochastic method, on the basis of random analysis, this could then be applied to the deterministic discussion of the entraining mechanism.

The first approach to this problem, by Townsend,[50] dealt with the wake behind a cylinder, establishing the two-component turbulent structure, as mentioned previously. Later work[51,52] took up this problem, both experimentally and theoretically, in two basic ways: the Eulerian method, i.e., conditional random analysis dealing with the discernment of turbulent or nonturbulent intervals by introducing the intermittency function on the record of the time series for the velocity fluctuation; and the Lagrangian method, dealing with the modeling of the eddy motion by a visualization technique. The Eulerian method has been applied to the analysis of heat transport in a two-dimensional heat jet. The application of conditional sampling to the problem of heat and momentum diffusion led to the conclusion that the temperature field is wider than the current field, in general, but the heat transport due to turbulence itself does not exceed the momentum transport.[53,54] The method has also been applied to the time series of velocity and temperature fluctuations and the Reynolds stress.[55] An investigation of the entraining mechanism on the basis of the duration characteristics of the turbulent state came to the conclusion that, in the outer part of the turbulent boundary layer, the largest velocity loss for the turbulent state occurs over a long period in the outer part of the turbulent boundary layer. This would mean that the Reynolds stress during that time might induce the entrainment of the irrotational cold fluid.

Of the studies of heat transport in the wake behind a plate,[57–59] particular mention should be made of an investigation by La Rue and Libby using spectral analysis[59]; it was found that the spectral profile of the intermittency function is in agreement with the Poisson-distribution power law, and the temperature fluctuation in the low-frequency range is closely related to the intermittency function. The structure near the interface is also related to the intermittancy function, and its characteristics are similar to those of the eddy couple, as shown by Grant,[60] who applied conditional random analysis in the established turbulent zone of a horizontal buoyant jet and proposed a theoretical diagram for the probability distribution of the velocity fluctuation (Figure 8). La Rue and Libby confirmed that the characteristics of random time series at the interface take on a two-component turbulent structure, comprising relatively slow motion by large-scale eddies and intense turbulence of short duration due to small-scale eddies. In addition,

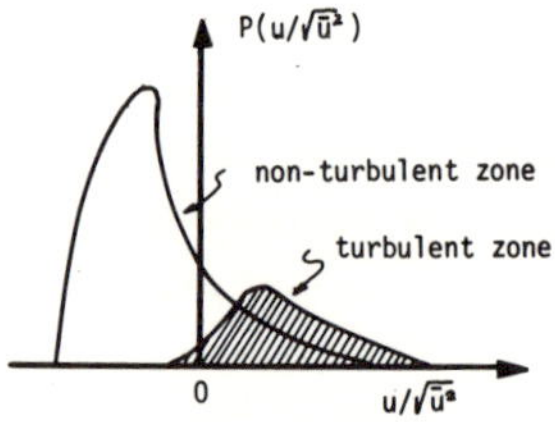

Figure 8 Probability distribution of the velocity fluctuation.[60]

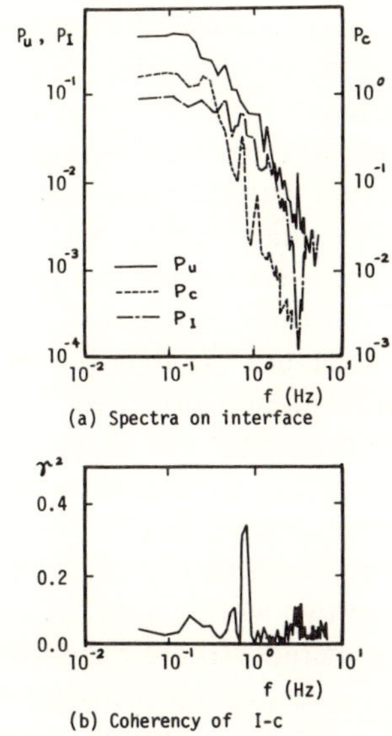

Figure 9 Statistical characteristics of turbulence: (a) spectra at interface; (b) coherence of *I–c*.

with regard to the spectra of velocity fluctuations, concentration fluctuations and the intermittency function denoted by P_u, P_c and P_I, respectively, in Figure 9a—it is found that the P_u spectrum seems to follow the $-5/3$ power law on the whole, whereas the P_c spectrum has clear periodicity. This discrepancy seems unreasonable on the view that the concentration fluctuation is induced by the velocity fluctuation. However, the P_I spectrum is in good agreement with P_c as regards the periodicity, and this suggests that the concentration fluctuation is induced by the coherent eddy structure developed at the interface; hence it is possible to sample the eddy scale contributing to the water-quality mixing.

It is more effective to introduce the coherence γ^2, which is computed from the cross-spectrum between the intermittency function I and the concentration fluctuation c. In Figure 9b, the coherence has a particular peak at $f = 0.75$ Hz but shows little response at $f = 0.25$ Hz and $f = 1.4$ Hz, which are dominant frequencies both for the periodicity of I and for the response in P_c. It is especially interesting that the frequency 0.75 Hz coincides with the Brunt–Väisälä number N, because it is supposed that the concentration fluctuation is able to follow the Väisälä oscillation number; this occurs under the dynamic equilibrium between the buoyancy and the restoring force of gravity.

In an investigation of the structure of large-scale eddies in the mixing layer by the Lagrangian method; on the basis of visual observation, the sweeping period and the Strohal number for large-scale eddies were determined and the relation of these quantities to the Eulerian spectra was established.[61,62] It was pointed out that the mixing layer develops by repeated pairing and bursting, whereas the Reynolds stress is a result, not of the bursting, but of interaction with the large-scale eddies.

There have been several studies dealing with the relation between the behavior of large-scale eddies in a jet and the geometric properties of the outlet.[63–65] In the case of a two-dimensional wall jet, Bajura and Catalano noted that the turbulent structure expands the mixing layer through the following process: (1) the formation of discrete eddies; (2) pairing of the eddies; (3) a lifting-off process from the jet mixing body to the ambient body; (4) dispersion of the eddies by the three-dimensional motion; (5) a laminarizing process until the next pairing.[65]

Examples dealing with the densimetric effect on the structure of large-scale eddies are presented in two works: a study of the relation between the density difference and the coherent structure controlling the development of the mixing layer[52]; and an investigation of vortex amalgamation in a two-dimensional horizontal jet in the case of turbid water, including a discussion of the relation between entrainment and large-scale eddies on the basis of a physical model of a couple of vortices with a scale of the order of the spreading width.[66] Although the results of these two studies did not agree, it is interesting that both linked entrainment with the phenomenon of entanglement.

Conclusion

In view of the increase of thermal wastewater coupled with the rapid increase in electric-power demand, the continuous monitoring of changes in the water environment is necessary. Two basic engineering systems are used for the treatment of thermal wastewater from a power plant: the cooling tower and the surface discharge. Both have merits and demerits, but surface discharge seems to have lower cost in general, taking account of the large size of power plants. However, as such a discharge affects a broad region of water, accurate and careful planning is necessary in the design of the effluent works and the prediction of water temperatures. Numerical modeling is commonly used for this purpose today, although some of the terms in the fundamental equations are not always correct for the simulation of the actual phenomenon.

In the last twenty years there has been remarkable progress in research on buoyant jets, and its results have been successfully reflected in numerical modeling. As the problems involved are three-dimensional, it is necessary to devise new computational techniques for the accurate prediction of buoyant-jet behavior. Another physical problem that remains to be solved concerns the hydraulic behavior in the water body receiving the thermal wastewater. This problem is more difficult in the case of the ocean, because it is not easy to establish the effect of the variation in tidal current and tidal water level on the advection and mixing of the effluent water body.

Further work is also necessary on entrainment in the near field, which is described at present in terms of the entrainment constant. While the concept of the entrainment constant has been useful in developing numerical models, it cannot properly be used in discussing the specific mechanism of entrainment.

Finally, there is good reason to hope that the concept of intermittency, while not new, will soon yield valuable insights into the solution of real problems.

ACKNOWLEDGMENTS

The authors would like to express their special thanks to Dr T. Goda, Head of the Water and Soil Environment Division of the National Institute for Environmental Studies, for his guidance in the writing of this review. The authors also wish to acknowledge the technical and other advice of Prof. A. Murota, of the Department of Civil Engineering, Osaka University.

REFERENCES

1. W. Tollmien, "Berechnung turbulenter Ausbreitungsvorgänge," *Z.A.M.M.*, **6** (1926).
2. H. Görtler, "Berechnung von Aufgaben freien Turbulenz auf Grund eines neuen Näherungansatzes, *Z.A.M.M.*, **22**, No. 5 (1942).
3. M. L. Albertson, Y. B. Dai, R. A. Jensen and H. Rouse, "Diffusion of submerged jets," *Proc. A.S.C.E.*, **74** (1948).
4. H. Lamb, *Hydrodynamics*, 6th ed. (Cambridge University Press, 1932) pp. 373–374.
5. G. I. Taylor, "Effect of variation of density on the stability of superposed streams of fluid," *Proc. Roy. Soc. A*, **132**, 499–523 (1931).
6. S. Goldstein, "On the stability of superposed streams of fluids of different densities," *Proc. Roy. Soc. A*, **132**, 524–548 (1931).
7. G. H. Keulegan, "Interfacial instability and mixing in stratified flows," *J. Res., Nat. Bureau Stand.*, **43**, 487–500 (1944).
8. B. R. Morton, G. I. Taylor and J. S. Turner, "Turbulent gravitational convection from maintained and instantaneous sources," *Proc. Roy. Soc. A*, **234**, 1–23 (1956).
9. T. H. Ellison and J. S. Turner, "Turbulent entrainment in stratified flows," *J. Fluid Mech.*, **6**, 423–448 (1969).
10. G. Abraham, "Horizontal jets in stagnant field of other density," *Proc. A.S.C.E.*, **91**, No. HY4, 138–154 (1965).
11. L. N. Fan and N. H. Brooks, *Proc. A.S.C.E.*, **92**, No. HY2, 423–429 (1966).
12. D. G. Fox, "Forced plume in a stably stratified fluid," *J. Geophys. Res.*, **75**, 6818–6835 (1970).
13. H. O. Anwar, "Measurements on horizontal buoyant jet in calm ambient fluid, with theory based on variable coefficient of entrainment determined experimentally," *La Houille Blanche*, No. 4 (1972).
14. P. R. Slawson and G. T. Csanady, "On the mean path of buoyant, bent-over chimney plumes," *J. Fluid Mech.*, **28**, part 2, 311–322 (1967).

15. L. N. Fan, *Turbulent Buoyant Jets into Stratified or Flowing Ambient Fluids*, Report No. KH-R-15 (W. M. Keck Laboratory of Hydrology and Water Research, C.I.T., 1967).
16. N. Shuto, "Buoyant plume in a cross stream," *Coastal Eng. Jap.*, **14**, 163–173 (1971).
17. V. H. Chu and M. B. Goldberg, "Buoyant forced plumes in cross flow," *Proc. A.S.C.E.*, **100**, No. HY9, 1203–1214 (1974).
18. D. R. F. Harleman, "Heat disposal in water environment," *Proc. A.S.C.E.*, **101**, No. HY9, 1120–1138 (1975).
19. K. D. Stolzenbach and D. R. F. Harleman, *An Analytical and Experimental Investigation of Surface Discharges of Heated Water*, Technical Report No. 135 (R. M. Parsons Laboratory for Water Research and Hydrodynamics, M.I.T., 1971).
20. T. Hayashi, M. Miyahara and M. Arita, "A mathematical model on the spread of heated water discharged horizontally at the water surface," in *U.S.–Japanese Seminar on Engineering and Environmental Aspects of Waste Heat Disposal* (1974).
21. H. Stefan and P. Vaidyaraman, "Jet type model for the three-dimensional thermal plume in a cross current and under wind," *Water Res. Res.*, **8**, No. 4, 998–1014 (1972).
22. E. D. Prych, *A Warm Water Effluent Analysed as a Buoyant Surface Jet*, Serie Hydrologi No. 21 (Sveriges Materologiska och Hydrolofiska Institut, Stockholm, 1972).
23. G. N. Abramovich, *The Theory of a Turbulent Jet* (M.I.T. Press, 1963).
24. D. Wada, "Study of thermal diffusion in a two layer sea caused by outfall of cooling water, in *Proceedings of an International Symposium on Stratified Flows* (Novosibirsk, 1972).
25. D. Wada, N. Katano, M. Kadoyu and H. Arai, *Study on Adaptability of Prediction Method of Simulation Analysis for Diffusion of Discharged Warm Water in the Sea*, Technical Report C73001 [in Japanese] (Central Research Institute of Electric Power Industry, Tokyo, 1975).
26. R. Frankel and J. Cumming, "Turbulent mixing phenomena of ocean outfalls," *Proc. A.S.C.E.*, **91**, No. SA2, 33–57 (1965).
27. Y. Jen, R. L. Wiegel and I. Mobarek, "Surface discharge of horizontal warm-water jet," *Proc. A.S.C.E.*, **92**, No. PO2, 1–28 (1966).
28. I. R. Wood and D. R. Wilkinson, "Discussion of 'Surface discharge of horizontal warm-water jet,'" *Proc. A.S.C.E.*, **93**, No. PO1, 149–151 (1967).
29. T. Hayashi and N. Shuto, "Diffusion of warm cooling water discharged from a power planet," in *Abstracts of Coastal Engineering Conference*, Session AV1 (The Institute of Civil Engineers, 1968) pp. 13–16.
30. H. Stefan and F. R. Schiebe, *Experimental Study of Warm Water Flow into Impoundment*, Parts 1–3, Minnesota University Report No. 101–103 (1968).
31. N. Tamai, R. L. Wiegel and F. F. Tornberg, "Horizontal surface discharge of warm water jets," *Proc. A.S.C.E.*, **95**, No. PO2, 253–276 (1969).
32. F. Engelund and F. B. Pederson, "Surface jet at small Richardson numbers," *Proc. A.S.C.E.*, **99**, No. HY3, 405–415 (1973).
33. N. Tamai, "Unified view of diffusion and dispersion in coastline waters," *J. Fac. Eng., Tokyo Univ.*, **31**, No. 4 (1972).
34. R. C. Y. Koh, "Two-dimensional surface warm jets," *Proc. A.S.C.E.*, **96**, No. HY6, 819–836 (1971).
35. H. Stefan, "Dilution of buoyant two-dimensional surface discharges," *Proc. A.S.C.E.*, **98**, No. HY1, 71–86 (1972).
36. R. Dornhelm, M. Nouel and R. L. Wiegel, "Velocity and temperature in buoyant surface jet," *Proc. A.S.C.E.*, **98**, No. PO1, 29–47 (1972).
37. K. Muraoka and K. Nakatsuji, "Entrainment constants in three-dimensional surface buoyant jet," *Tech. Rep. Osaka Univ.*, **27**, No. 1389 537–545 (1977).
38. S. Corrsin, *Investigation of Flow in an Axially Symmetrical Heated Jet of Air*, Wartime Report W-94 (National Advisory Committee for Aeronautics, 1943).
39. W. D. Baines, "Discussion of 'Diffusion of submerged jets,'" *Trans. A.S.C.E.*, **115**, 677–684 (1950).
40. D. Wada and N. Katano, "A study of diffusion in connection with the intake and disposal of industrial water," in *Proceedings of the Fifteenth Japanese Conference on Coastal Engineering* [in Japanese] (1968) pp. 297–306.

41. O. M. Phillips, The Dynamics of the Upper Ocean (Cambridge University Press, 1966).
42. R. Bolgiano, Jr., "Structure of turbulence in a stably stratified medium," *J. Geophys. Res.*, **62** (1962).
43. J. L. Lumley, "The spectrum of nearly inertial turbulence in a stably stratified fluid," *J. Atmos. Sci.*, **21**, (1964).
44. T. Iwasaki and T. Abe, "Turbulence in the weakly mixing zone of the interface," *Coastal Eng. Jap.*, **15**, 141–151 (1972).
45. T. Tsubaki and T. Komatsu, "Spectral properties of turbulence and density fluctuation in a stratified shearing flow," *Proc. A.S.C.E.*, No. 268, 63–74 (1977).
46. N. Tamai and O. Nishimura, "Study of internal wave and mixing properties in a stratified flow," in *Proceedings of the Seventeenth Japanese Conference on Hydraulics* [in Japanese] (1973) pp. 32–37.
47. P. G. Drazin and L. N. Howard, "Stability in a continuously stratified fluid," *Proc. A.S.C.E.*, **87**, No. EM6, 101–116 (1961).
48. K. Muraoka and K. Nakatsuji, "Measurements of turbulence quantities in three-dimensional surface buoyant jet," *Tech. Rep. Osaka Univ.*, **28**, No. 1425, 281–289 (1978).
49. J. Lauter, "New trends in experimental turbulence research," *Ann. Rev. Fluid Mech.*, **7**, 307–326 (1975).
50. D. D. Townsend, *The Structure of Turbulent Shear Flow* (Cambridge University Press, 1956).
51. S. J. Kline, W. C. Reynolds, F. A. Schraub and P. W. Runstandler, "The structure of a turbulent boundary layer," *J. Fluid Mech.*, **30**, 741–773 (1967),
52. G. Brown and A. Roshko, "On density effects and large structure in turbulent mixing layers," *J. Fluid Mech.*, **64**, 775–816 (1974).
53. D. E. Davis, J. F. Keffer and W. D. Baines, "Spread of a heated plane turbulent jet," *Phys. Fluids*, **18**, No. 7, 770–775 (1975).
54. P. E. Jenkins and V. W. Goldschmidt, "Conditional (point averaged) temperature and velocities in a heated turbulent plane jet," *Phys. Fluids*, **19**, No. 5, 613–617 (1976).
55. R. A. Antonio, A. Prabhu and S. E. Stephenson, "Conditional sampled measurements in a heated turbulent jet," *J. Fluid Mech.*, **72**, 455–481 (1975).
56. R. A. Antonio, "The Distribution of Temperature in the Intermittent Region of a Turbulent Shear Flow, in *Fifth International Heat Transfer Conference* (1974) pp. 95–98.
57. R. M. Thomas, "Conditional sampling and other measurement in a plane turbulent wake," *J. Fluid Mech.*, **57**, 540–582 (1973).
58. L. S. G. Kovasznay and S. F. Ali, "Structure of the Turbulence in the wake of a heated flat plate," in *Fifth International Heat Transfer Conference* (1974) pp. 99–103.
59. J. C. La Rue and R. A. Libby, "Temperature and intermittency in the turbulent wake of a heated cylinder," *Phys. Fluids*, **17**, 873–878 (1974).
60. H. L. Grant, "The large eddies of turbulent motion," *J. Fluid Mech.*, **4**, 149–190 (1958).
61. C. D. Winant and F. K. Browand, "Vortex pairing: the mechanism of turbulent mixing layer growth at moderate Reynolds number," *J. Fluid Mech.*, **63**, 237–255 (1974).
62. F. K. Browand and P. D. Weidman, "Large scale in the developing mixing layer," *J. Fluid Mech.*, **76**, 127–144 (1976).
63. H. D. Becker and T. A. Massaro, "Vortex evolution in a round jet," *J. Fluid Mech.*, **31**, 435–448 (1968).
64. G. S. Beavers and T. A. Wilson, "Vortex growth in jets," *J. Fluid Mech.*, **44**, 97–112 (1970).
65. R. D. Bajura and M. R. Catalano, "Transition in a two-dimensional plane wall jet," *J. Fluid Mech.*, **70**, 773–799 (1975).
66. H. Kikkawa, S. Fukuoka and K. Yoshikawa, "A study of fluid–solid particle interaction," *Proc. A.S.C.E.*, No. 260, 89–100 (1977).

Kohji Muraoka
National Institute for Environmental Studies, Japan
Keiji Nakatsuji
Osaka University

INFLUENCE OF FOREST VEGETATION AND AGRICULTURE ON THE ACIDITY OF FRESH WATER

Introduction

For several years the authorities in Norway and Sweden, and also, to some extent, those in Canada and the U.S.A., have been concerned with both the short- and the long-term effects of acid precipitation caused by an increased combustion of sulfur-bearing fuels or produced in the metallurgical industry. The regional pollution and acidification of rain and snow and also of atmospheric dry deposition in the industrialized world can no longer be disputed.

Locally, as in the Sudbury area of southern Ontario, Canada,[1] the emission of acid pollutants and precipitation may be extreme. In the Sudbury area, harmful effects on the vegetation, with subsequent soil erosion, and also a certain impact on the freshwater pH seem to have been proven. Harmful effects of acid precipitation on freshwater streams have also been assumed for other areas, for example, for southern Scandinavia in the S.N.S.F. project report.[2] However, the same report states explicitly that no harmful effect on the foest could be attributed to the acidity of the rain.

The Norwegian S.N.S.F. project has carried out two large taxation studies on the growth of the coniferous forest in southern Norway and compared the results with the inner parts of east Norway.[2] Both studies involve a very large amount of material and seem to have been carried out with great care.

The summary of the report includes the following statements on tree growth in the two regions.

> Regions assumed to differ in exposure to acid deposition have been compared with regard to past diameter growth. So far no effects have been found that can be related to acid precipitation, nor have effects been detected on sites supposed to be most sensitive to acidification.

> The studies on height growth in field experiments have not revealed any negative effects of acidification on Norway spruce in a forest nursery.

The marked difference (amounting to contradiction) between the Canadian and Norwegian results seems to reflect the great difference in the concentration of the emission and in the precipitation of sulfates and free protons (acids) in the two studies. The total annual precipitation of sulfur compounds (not only H_2SO_4) falling in the whole of Norway (324,000 km^2) corresponds to only one-third of the localized emission of SO_2 from the smelters of the Sudbury nickel mines ($\sim$2 million tons).

The SO_4^{2-} content of lakes in southern Norway is less than one-third or one-fifth of that found in various lakes at Sudbury.[1]

The precipitation at Sudbury may be as high as 5.5 g SO_4^{2-} per m^2 in 28 days, and the rain may have pH values as low as 2.85.[3]

The summary reports of the O.E.C.D. programme on the long-range transport of air pollutants[4] have shown that the wet deposition of acid in 1974 in the most exposed locality of sourthern Norway contained up to 90 meq acids per m^2. Normally the values are considerably lower. The large amounts of acids found in the southern coastal area of Norway are mainly due to the high levels of precipitation in the region, rather than to the specific acidity of the rainwater.

The average concentration of sulfate in the rain and snow in southern Norway in 1972–1975 was only of the order of 2–3 mg SO_4^{2-} per liter, and only part of this was present as corresponding H_2SO_4. A considerable proportion of the *acids* in snow and rain is, however, precipitated during limited periods called "episodes," and the rainwater may then reach pH values of slightly below 4. The weighted-average pH of the precipitation, based on data from July, 1972, to June, 1975,[4] is 4.5 ± 0.3 in southern Norway. Calculating the proton concentration in the rain arbitrarily as if it originated from sulfuric acid, values corresponding to between 1 and 3 tons of H_2SO_4 per km^2 per year are obtained.

In the southern coast area, where acidic river water is most common and the average precipitation may have pH values as low as 4.2, the chlorine content often dominates over sulfate. As the electrolytes are fully dissociated, it would then be as correct to speak of a hydrochloric-acid content as to speak of a sulfuric-acid content. Even if the total H^+ concentration in the precipitation were attributed to sulfuric acid, the SO_4^{2-} content would still be higher than would correspond to the proton content. Thus, a considerable proportion of the sulfate is already neutralized by cations other than H^+.

Table I shows the average composition of the rainwater in the extreme south of Norway, corresponding to a pH value of 4.3. The total cation content for the figures in Table I is 440 μeq/liter, while the total anion content is 364 μeq/liter; the discrepancy between these figures may be due to nitrates, to carbonates in periods with pH > 5 or to dust particles.

Neutral salts exceed the content of acids even in the most acid areas of

TABLE I

Average rainwater composition, μeq/liter, for the extreme south of Norway

H^+	Na^+	K^+	Mg^{2+}	Ca^{2+}	SO_4^{2-}	Cl^-
50	265	11	62	52	72	295

southern Scandinavia. This is important in the analysis of the water geochemistry of a catchment area. The chlorine and sulfate content and their mutual ratios may vary considerably in the Scandinavian precipitation and only part of the sulfates can be attributed to the anthropogenic acid emissions. There is, however, a fair uniformity in the average acid content in the rain and snow precipitation of southern Norway. Usually the pH varies from around 4.3 in the coastal areas of south-eastern Norway to 4.6 inland, i.e., from 50×10^{-6} to 25×10^{-6} mol. H_3O^+ per liter. The content of electropositive ions—Na, Mg, Ca, K, etc.—may reach 500 μeq/liter. The amount of rain, however, varies considerably more.

As mentioned, the S.N.S.F. Acid Precipitation Project did not find any harmful effect on forests which could be attributed to the acidity of the rain.[2] However, it was stated in the report that there is a direct relationship between the episodic input of atmospheric acids and the pH of stream waters. This statement has subsequently been challenged. In addition, it was assumed, without further examination, that the acidity of the precipitation is responsible for the acidity of various rivers and lakes, and that the fish population has been strongly diminished or entirely wiped out, presumably as a result of the low pH in the precipitation water.

The hypothesis that the fish mortality is due to the acid precipitation has been repeated in a large number of articles in the international literature. In many countries this idea seems to have gained general acceptance. Even this hypothesis has recently been challenged, in a monograph which demonstrates the importance of vegetation and humus as sources of acidification.[5]

Drawing attention to cultural changes in the vegetation, the monograph reports laboratory experiments and field observations which demonstrate that ion-exchange reactions in the raw humus layer of podzol profiles and the influence of cation uptake by plant growth are the most important factors in determining the pH of runoff water. Noting the increase in biomass and subfossil humus due to changes in agriculture, cattle farming and forestry during the last century, the monograph concludes that most catchment areas represent increasing proton reservoirs.

The proton content of the upper soil profiles of these reservoirs is flexible, representing low-pH buffering systems in the water chemistry. As early as 1903, Helland reported that lakes and tarns in South Norway had lost their fish population in around 1890, probably as a result of acidic water.[6]

Protons are supplied to these reservoirs by precipitation. At the same time neutral electrolytes are added, mainly from natural sources.

Protons are *produced* in the reservoir as a result of "proton pumping" by the living plants, with the formation of humic acids.

Protons are *consumed* in the reservoir by weathering reactions attacking the primary minerals underground.

In certain periods proton may be *accumulated* in the reservoir.

The outflowing water in the brooks and rivers of the water courses may *extract* protons from the reservoir.

A diagram of the ecosystem is shown in Figure 1.

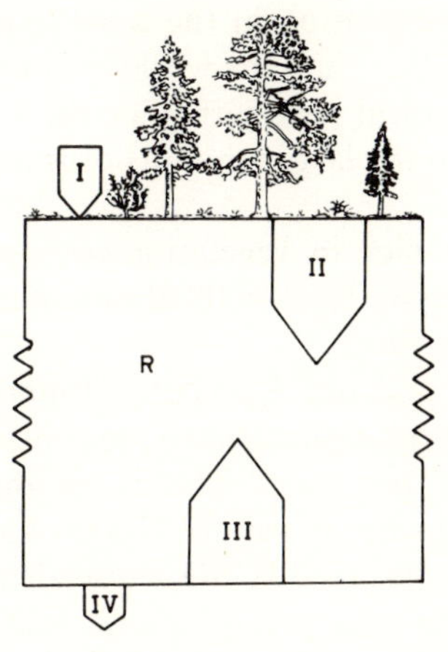

Figure 1 Reservoir ecosystem: (R) Proton reservoir; (I) Protons supplied by precipitation; (II) Protons produced within reservoir; (III) Protons used or accumulated; (IV) Proton outflow.

Natural Processes

The idea that a direct relationship exists between the pH of precipitation falling in a given watershed and the pH of the outflow waters rests on the belief that acids are produced and consumed in the catchment area in considerably smaller amounts than those brought in by the precipitation. In the past, this belief could be sustained because ignorance of soil chemistry was widespread and quantitative data dealing with the natural production of natural acids were sparse and were mostly not directly applicable in any specific case.

Concerning the consumption of natural and anthropogenic acids, it is generally agreed that carbonate rocks are able to neutralize any amounts of acidic water that may exist in practice. In areas where the bedrock consists of acid (quartz-bearing) silicate rocks as gneisses, granites or quartz diorites, it has often been assumed that the proton consumption due to weathering is of little importance in boreal and subarctic areas. This is a myth, however, and has no foundation in fact.

In the alpine regions of the Cascade Mountains in the U.S.A., Reynolds and Johnson found that the breakdown of minerals from the quartz diorites due to weathering corresponds to a proton consumption of 0.3 meq per liter of precipitation.[7] For an annual rain- and snowfall of 3000 mm water per year, this corresponds to the liberation of 930 meq electropositive monovalent and bivalent cations per m^2 per year. This is more than 10 times as much as the acid precipitation in the most exposed areas of southern Norway, and it is also higher than the world average.

The high rate of chemical weathering in the Cascade Mountains can probably be attributed to the large amounts of fresh rock material exposed by recent glaciation.

Studies of podzol profiles in vegetation-covered areas in the Central Norwegian Mountain region have enabled minimal values to be calculated for the postglacial weathering.

In the A horizon of podzols, Rueslåtten found that the liberation of electropositive ions has been equivalent to the consumption of at least 1 ton of sulfuric acid per km^2 per year.[8] This is an absolute minimum. If we include the weathering in the B part of the profiles, the weathering data correspond to 4 tons of consumed H_2SO_4 equivalents per km^2 per year.

Analysis of river water and precipitation[9] leads to higher weathering rates in similar areas, corresponding to 7 tons of sulfuric acid per km^2 per year; this gives a figure of 140 meq per m^2 per year, which is obviously much higher than the acid precipitation in most parts of the world (excluding extreme cases like the Sudbury mining and smelter region). Therefore there are strong indications that the natural production of acids as a result of plant growth must be higher than the anthropogenic component from air-borne pollutants. Since the production and consumption of acids normally take place simultaneously in the same area, quantitative evaluation of the individual factors is extremely difficult.

In a few areas of the world, however, it is possible to assume that neutralization reactions make only a small contribution to the total turnover of acids. Geologically, such areas are those which consist of deeply weathered rock formations. In such areas the primary minerals have already been decomposed and all the strongly electropositive ions have been removed.

In parts of the Amazon Basin, the uppermost rock profiles consist mainly of kaolin (an aluminum silicate) and gibbsite and other hydroxides of aluminum and iron. At the same time the topography of these regions is relatively flat, and only small amounts of the precipitation water will therefore penetrate sufficiently deeply through the profiles to react with the underlying primary silicate minerals that contain electropositive ions. Most of the precipitation water—except the evapotranspiration component runs off to the watercourses without any neutralization of its acid-proton content, regardless of whether the protons originated from precipitation or are due to acid production in the catchment area.

Good examples of such drainage conditions are found in Amazonia. Soli reports that, where the relief of the headwater zones of Amazonian rivers is relatively flat, there are "blackwater" streams with a pH of 3.8–4.9 and "clearwater" streams with a pH of 4.5–7.8 but, where the relief is rugged, with mountain slopes, there are "whitewater" streams with a pH of 6.2–7.2.[10] The difference between the blackwaters and the clearwaters is that the very acid blackwater originates from areas characteristic of tropical lowland podzols, whereas the clearwater drains from other soil types.

In a private communication of November 8, 1977, Dr George Irion of the Institute für Meeresforschung, Wilhelmshafen, West Germany, who has worked extensively in the German Amazon project, points out that the suspended minerals in the Rio Negros consist solely of kaolinite and traces of gibbsite.

H. Klinge and W. Ohle give even lower values for the pH of some Amazon blackwater rivers.[11] They have recorded a pH of 3.58.

As the Amazon rivers drain areas where the influence of anthropogenic air pollution is minimal and the neutralization effect from the primary minerals is low, the low pH of the river water can only be attributed to the effect of *in situ* reactions between the ground material and the precipitation water.

At the beginning of the nineteenth century, Alexander von Humbolt noticed that these rivers had a peculiar fauna, characterized by a lack of crocodiles and mussels with calcareous shells and by a low fish population. There is no possibility of industrial pollution in these tropical areas at that period in history.

Similar low pH values are characteristic of lakes in the Norwegian peat districts.[12] In the mountain regions of southern Norway, *Sphagnum* peat moors may have porewater with a pH of about 3.0—more than 10 times the acidity in acid rainwater.[5]

It has often been stated that these low pH values in *Sphagnum* peat and peat tarns are due to the so-called humic acids.

However, even superhumic water with a dark brown coloration contains only about 20–50 mg dissolved carbon per liter. Even if all this carbon had been present in the form of the strongest known natural acid (oxalic acid), the pH value would only be around 4.0, and all other humic acids have aciditics considerably lower than oxalic acid. The exaplanation for the low pH values is found in the "proton-pumping effect" of the *Sphagnum*.[13,14]

Sphagnum and all other mosses have no root system; they take their necessary cations from the precipitation. They are synthesizing an insoluble cellulose–uronic-acid–pentosane aggregate. Uronic acids make up 15–30% of the dry matter. These uronic acids act as ion exchangers: they react with dissolved salt—as Na^+ and Cl^-—in the rain and aerosols, forming sodium uranate and free hydrochloric acid.

The acidification effect of such biological ion-exchange systems depends on two factors: (1) the annual biological production of ion exchanger; and (2) the quantity of electrolytes falling as rain or dry deposits on this carpet ion exchanger. The annual production of dry matter in normal *Sphagnum* peats in Norway corresponds to 50–120 g of polyuronic acids per m^2 per year. The equivalent weight of uronic acid is 178. The bioproduction thus corresponds to 280–710 meq H^+ per m^2 per year, which, in turn, corresponds to a maximum potential production of acids equivalent to 14–35 tons of H_2SO_4 per km^2 per year. This potential is relatively uninteresting, because there is no area where the precipitation of salts will reach this potential. It means, however, that under given conditions, bioproduction is able to convert neutral salts from the precipitation to mineral acids by a simple ion-exchange reaction.

In the southernmost part of Norway, where acid lakes and brooks are common, the average precipitation has a pH of ~4.3 and contains ~0.4 meq electropositive ions per liter. Assuming complete cation exchange with organic insoluble cation-exchange material in the hydronium state, these ion concentrations would result in runoff water of pH 3.5. If, therefore, the runoff water should happen to have a pH value of 4.3, i.e., the same value as the precipitation, this does not mean that it is the acids from the rain which are found in the rivers.

Because of the evapotranspiration, the concentration of salts will increase in the soil water, especially during dry periods, and a subsequent rainy period would then flush out waters of even lower pH, provided no neutralization reactions take place.

Such reactions are of different types. There may be weathering reactions with minerals containing electropositive ions and also the reactivation of such ions from decaying dead plant material which is mineralized. Furthermore, the reduction of sulfates consumes protons. This reaction may be written as follows:

$$SO_4^{2-} + 2CH_2O + 2H^+ = H_2S + CO_2 + 2H_2O;$$

CH_2O represents organic matter. In other cases the equation is written:

$$SO_4^{2-} + 2H^+ + 4H_2 = H_2S + 4H_2O.$$

These crude overall equations do not permit calculations of the necessary partial oxygen pressure for equilibrium between H_2S and SO_4^{2-}, nor of the Eh/pH relations for the reaction. At low pH this reaction takes place according to the following equations:

$$SO_4^{2-} + 10H^+ + 8e^- = H_2S + 4H_2O;$$

$$\mathrm{Eh} = 0.303 - 0.074\mathrm{pH} + 0.0074 \log \frac{[SO_4^{2-}]}{[H_2S]}.$$

For equal activities of SO_4^{2-} and H_2S, the last term is zero. We may, therefore, draw a line in an Eh–pH diagram, starting at pH = 0 and Eh = 0.303 V, and dropping 74 mV for every pH unit. At all values below this line, $[H_2S]$ dominates over $[SO_4^{2-}]$.

Thus, at low Eh, the reduction of sulfate to sulfide implies a consumption of 10 H^+ ions and 8 electrons. The 8 electrons are produced by oxidation of two organic carbon atoms. As long as the material is protected from oxidation, the sulfides remain in the sediment. By oxidation during a dry period, new sulfates and acids may be formed, resulting in an acid shock in the runoff water in a new rainy period.

The ability to synthesize insoluble ion-exchange material is not restricted to mosses. In most humic material ion-exchange material in the proton (hydronium) state is present. This is especially the case for the raw humus formed on heathland and in coniferous forest. Bergseth has examined 185 localities in Norwegian and Swedish forest soils of fairly homogeneous soil type, i.e., podzols and brown earth.[15] All the observations have demonstrated that there is a close relationship between the type of vegetation and the pH value of the topsoil and also the quantity of exchangeable protons in the upper—O, A and B—horizons of the soil profiles. However, in the deeper parts—at a depth of 60–80 cm in the C horizon—the pH did not vary significantly from one type of vegetation to another. In areas, with deciduous trees and an undergrowth of grasses, the pH value in the top soils varied from 5 to 6.5, while the exchangeable proton content in the topsoil was about 20 meq per 100 grams of soil. In areas with coniferous forests and heather or lichens in the undergrowth, pH values between 3.5 and 4.5 are normal.

To a certain extent these values show the influence of the deep root systems in the deciduous forests as compared to the spruce forests. Thus the hardwood vegetation is bringing up cations from the subsoil and depositing them on the topsoil. The spruce, however, recirculates the ions from the litter to a much higher degree. It is not the case that the annual base uptake is much less in hardwood (leafwood) than in softwood. In hardwood soils the worm fauna are also much more active in circulating mineral components and bringing bases up from the deeper strata.

Although there are large differences in different environments, all soil-forming processes in humid climates sooner or later lead to acidification of soils on silicate bedrock, unless countermeasures are taken. This may be done by the addition of limestone or other bases by man. The acidification of the topsoil as a consequence of vegetation is increasing, and affects the rate of chemical weathering of minerals underground. These processes are stages in the geochemical circulation of elements.[16]

The natural acidification of the soils due to plant growth and coaction between neutral salts in the precipitation and ion-exchange systems in the subfossil humic have gone on ever since plants invaded the dry earth in the Late Silurian epoch. Had it not been for the erosion of the weathered crusts and the steady exposure of new unweathered rock material to the soil-forming processes, biological excreta would long ago have "poisoned" the Earth's crust and driven life back into the oceans. The present addition of anthropogenic acids to the geocirculation, however, has attracted more public interest than the natural processes.

Generally it has been postulated that the acidification of soils is equivalent to and a result of the quantity of H_2SO_4 and other strong-acid deposits on the ground. However, it should be noted that H_2SO_4 may be reduced to H_2S in wet soils and HNO_3 is reduced to NO_x and N_2, thus diminishing the acidifying effect of these acids. Sulfate ions will be absorbed by growing plants and used in the biosynthesis of sulfur-bearing amino acids, which are essential components of plant protein. After the death of the plants, these sulfur-bearing components will decompose and sulfur will circulate. There may either be direct and complete oxidation to sulfate ions, or else liberation of hydrogen sulfide, which again will be oxidized to SO_2 and H_2SO_4.

There has been much discussion as to what proportion of the world's atmospheric sulfate is due to circulating sulfur involved in the biocirculation system. Some scientists believe that as much as 87% of the total emission of sulfur components is due to natural processes and only 13% is of anthropogenic origin.[17] Of this 87%, 87% is supposed to originate from biological materials by decomposition processes, the rest coming from ocean spray and volcanoes.

In itself it is of no interest to know what part of the sulfur in the higher atmosphere and in rain and snow is due to natural decomposition and what part is of anthropogenic origin: it will always be the local catchment area which is of interest. Even if the biogenic H_2S does not reach the air masses above the top of the vegetation before it is oxidized or absorbed and recirculated, it remains the case that there is an input and output of sulfur components in the ecosystem.

The opinion has been advanced that, if the total concentration of strong-acid anions in precipitation rises, the intensity of weathering and leaching in the soil will be increased. Wiklander opposes this viewpoint on the basis that, as a result of the interaction between mineral soil and acid deposition, the assumed equivalence is limited to soils with certain properties, and argues that the proponents of this view fail to take into account the significance of the "base" cations in the precipitation and the variation of the bonding strength of the H^+ ion.[16] In fact, the total chemical composition of the infiltrating precipitation water and the properties of the soil are of paramount significance in determining the cation-exchange effect and the actual soil acidification. Anions of strong acids—SO_4^{2-}, NO_3^- and Cl^-—have little or no influence in this connection. Their effect is chiefly on the ionic strength and thereby the ionic activity of all the ions in solution.

Another point which has generally been underestimated in Scandinavia is, as already mentioned, the rate of weathering and the liberation and addition of "base" cations: K^+, Na^+, Ca^{2+}, Mg^{2+}, etc. This underestimation seems to arise from calculations of soil profiles based on the assumption that quartz behaves more or less as an inert mineral. Thus, determinations of the ratio between minerals containing K^+, Na^+, Mg^{2+}, Ca^{2+}, etc., and quartz in the upper and deeper parts of soil profiles show a very small relative enrichment of quartz and, on this basis, low weathering rates were deduced.

However, analyses carried out in many parts of the world, including the subarctic regions, show that most stream waters contain between 10^{-4} and $10^{-4.5}$ mol. H_4SiO_4 per liter; this corresponds to quartz saturation and even higher values. For example, all streams entering Lake Mjøsa in Norway are supersaturated with free quartz.[5] As a result of the growth of diatom algae in the lake, the outflowing river water is slightly undersaturated, but even this water contains more than 10–100 times as much H_4SiO_4 as is found in the precipitation water.

Thus it is clear that water passing through and above geological deposits contains considerable concentrations of dissolved silica. The present rate of solution of quartz is not much less than that of K-, Na-, Ca- and Mg-bearing silicate minerals. This was also found to be the case in the A horizon of podzol profiles at Dagali in Norway at altitudes between 900 and 1100 m.[8] It was concluded that 46% of the quartz and 53–54% of the feldspar minerals had gone into solution since the profile started to develop. The enrichment of quartz relative to feldspar minerals in a soil profile is therefore very small.

On the other hand, dark minerals such as chlorite, trioctahedral micas and amphiboles have nearly completely disappeared in the A layer. If a H_4SiO_4 content in river water of as little as 1 mg/liter is assumed in an area with an annual runoff of 1000 mm and soil or rock material containing 60% SiO_2, this H_2SiO_4 content may be used to calculate an absolute minimum value for the rate of weathering.

Supposing that the rate at which minerals enter solution is proportional to their relative concentration in the regolith or bedrock and is the same as the rate for silica, it is found that in granitic areas the dissolved bases will be sufficient to neutralize one ton of H_2SO_4 per km^2 per year, for a H_4SiO_4 content of 1 mg/liter in the water and 1000 mm precipitation. However, most of the SiO_2 is reprecipitated in the soil and does not enter the streams. The weathering is therefore more than equivalent to the present precipitation of all cations, including the protons.

All biological processes will influence the soil acid. Not only is there indirect production of acids through ion-exchange reactions between neutral salts in the precipitation and the humic components formed from the litter of growing forests, but also the growing plants will themselves produce acids equivalent to the amount of base uptake from the roots. The ash content of wood from different plants is rather variable: the variation may be 0.25–5% of the dry weight of the wood.[18,19] For pinewood, which in Norway is the most important timber on the shallow and coarse-grained soil areas where the acidification of rivers is at its most severe, the ash content varies between 0.3 and 0.6%. CaO is the dominant base component of the ashes ($\sim$50%), while K_2O, MgO and Na_2O together make up about 30%. The anions SiO_2^{2-}, $P_2O_5^{3-}$, SO_4^{2-} and Cl together make up less than 15%.

In Scandinavia the production of wood varies considerably in the different areas, depending on altitude, soil quality and topography. In the most productive areas this range may be between 120 and 750 m^3 per km^2 per year. This corresponds to between 50 and 300 tons of dry wood and the amount of bases extracted from the area by this amount of wood may be 10–60 meq per m^2 per year. The extracted bases will therefore correspond to an acidification of 10–60 meq of acids per year per m^2. Again this corresponds to the equivalent of 0.5–3 tons of H_2SO_4 per km^2 per year. These figures are of the same order as the present precipitation of strong acids by rain and snow.

In young growing forests the amount of bases accumulating annually is considerably higher. In fertilized young forests the extracted bases may be many times higher than the figures given above.

Therefore, the intensive production of wood involves the extraction of bases and acidification; this can only be counteracted by bringing into the area large amounts of basic materials, such as $CaCO_3$.

Up to 1960 the timber was barked at the logging site before being brought out from the forest. In modern forestry, even the barks are removed from the forest area, and titration tests in the laboratory have shown that in such cases wood from unfertilized forest corresponds to an average of 200 ml of normal base per kg. That means that in areas where the production of timber is 250 tons per km^2 the trees are extracting bases equivalent to the present acid precipitation (pH 4.3) and 1000 mm rainfall per year.

In southern Norway logging has been practised in the productive forests, at a level more or less corresponding to the natural wood production, for more than 300 years. No bases have been added to the logging areas, and the productivity does not seem to have decreased. This means that weathering processes in these areas must have produced, annually, amounts of bases at least corresponding to the ash content, which in turn, corresponds to the present acid precipitation. However this was by logging methods which left in the forest the needles, twigs and bark, with their high base content; only the stem wood was removed.

Modern logging methods will, on average, more than double the extraction of K^+, Na^+, Ca^{2+} and Mg^{2+}, thus adding considerably to the natural acidification. Natural forests are very effective in the production of biomass on unfertilized ground. Whereas an annual production of 250 tons of grain per km^2 requires the extensive use of fertilizers, natural forests are able to produce this amount of dry biomass by utilizing ions from precipitation and from weathering processes.

In the natural preindustrial forests the trees died and rotted in the place where they had grown, and the bases were returned to circulation. This is the same as the situation on grass pastures: the grazing economy involves very small extractions of bases by milk, meat and wool, since nearly all the mineral components extracted from the ground by the grass are returned in the dung from the sheep and cattle. When pastures are reforested and the accumulated bases are exported, man has introduced acidification into the catchment area.

In nature at least the following five factors seem to act in the acidification of the soil profiles:

1) the uptake of bases by growing plants;
2) the formation of CO_2 in the ground;
3) oxidation of nitrogen and sulfur compounds in the ground;
4) the introduction of acid or acid-forming litter from vegetation;
5) the precipitation of acid components with rain.

Professor Wiklander in Sweden has recently dealt with these factors for acidification of the ground.[20] He pointed out the high content of cations in litter from elm, ash, alder, maple and oak and partly also birch. Needle litter from coniferous forests and peat moss have low base values. The highest value was found in elm leaf (250 meq per 100 g dry substance) and the lowest in raw humus (9 meq per 100 g); the figure obtained in moss peat was 13 meq per 100 g. If the organic litter is relatively rich in cations, as in cultivated plants and grass and many deciduous trees, the humic acids will be more or less base-saturated and will not be able to give acid runoff water by ion exchange with electrolytes in the precipitation.

On the other hand, the raw humus and needles from pine and spruce will react under ion-exchange processes and therefore represent acidifiers. A study of the effect of forest vegetation on the chemical status of the humus layers in Norwegian soil profiles has indicated the variation in the content of Ca, Mg, Mn, K, P, N, Si and ash.[18]

Further acidification in the ground results from the formation of CO_2 by the oxidation of organic material together with the liberation of CO_2 from plant roots and from microorganisms in the soil. In humid climates the CO_2 content will amount to several percent of the gas phase in the topsoil. The CO_2 production involves a considerable extracting effect on, for instance, $CaCO_3$ in the ground, and it will have a considerable influence as long as the pH is above 5.5. The leaching effect of CO_2 may reach some few tons of $CaCO_3$ per hectare per year. This corresponds to more than 100 tons of H_2SO_4 per square kilometer per year.

The nitrification of nitrogen components according to the equation $NH_4^+ + 2O_2 = 2H^+ + NO_3 + {}^-H_2O$ represents an acidification needing an addition of bases corresponding to $\sim$7 kg $CaCO_3$ per kg N in the precipitated or added ammonium. Hence, in all areas where nitrogen fertilizers are used, there will be very strong acidification. Even in unfertilized areas, the natural precipitation of ammonium components has an important acidifying effect on the soil. The same is true of nitrogen bound by mycorrhiza of the plant roots.

The acidification due to the oxidation of sulfide-rich soil in dry periods may often bring the pH below 3 in the groundwater.

Taken together, these factors will normally far outweigh the influence of added anthropogenic acids in most areas.

In natural soils we find all the stages of acidification. In areas with carbonate rocks and vegetation rich in calcium and other bases, the pH in the deeper parts of the soil profiles is normally above 7, and drainage water will, even under conditions of very acid precipitation, have high pH values. In podzols and acid brown earth the pH may be well below 4.0. As the degree of base saturation may be as low as 5% of exchangeable positions, these acid soils will act as buffers on the acid side. The result, nearly independent of the pH of the incoming solution but depending to some extent on the electrolyte content, is that the percolating water will become acid. Thus, there is usually a relationship between acidification of the soil and acidity of the runoff water.

By introducing salt solutions into the soil, the degree of base saturation will increase, whereas the runoff water will take with it parts of the proton content from the acid soil and thus give acid runoff water in the brooks and streams. Only a few seconds of contact time between the percolating solutions and the soil is sufficient to equilibrate these systems. On the other hand, very acid rainfall will result in outflowing water of lower acidity than the influx. This again will lead to a small increase in the acitity of the soil profile, and a decrease in the degree of base saturation.

As the exchange capacity in a few centimeters of soil is large, hundreds of years of precipitation of the present acidity may be necessary in order to introduce the same numbers of protons as are present in exchangeable positions in acid soil. Because of the weathering processes in the deeper parts of the profiles and the upward transport of cations through the roots and stems of the vegetation, new cations are added to the uppermost part of the profiles. This will reestablish the base-to-proton ratio, and this ratio will depend on the soil profile and on the type of vegetation.

The Influence of Man on Vegetation and Soil Acidity

The neolithic revolution is supposed to have started in the Middle East some 8000 years ago and slowly spread over most of the world. This transition from food collection and hunting to food production involved the hardest attack on virgin nature in the history of mankind. Forests were cleared and changed into pastures and fields. In southern Norway seminomadic cattle farming probably began some 3000 years ago and has developed from the pre-Roman Iron age until the present.[21] This left the country in a condition which differed chemically and biologically from its natural state.

In areas of grazing and farming the raw humus of the heathland was replaced by grass vegetation, and soil erosion became more intensive. Especially during the eighteenth and nineteenth centuries, many new marginal farms were taken into use. All the soil which could be tilled was used for food and fodder production. The forests were cleared and in nearly all possible pasture areas the natural vegetation was burned and the land was grazed. All plant products which could be used for food production were utilized. Grass and hay were brought in or grazed, even when they were far away from the home farm. The landscape became open and was marked with a mosaic of small farms and fields. The productivity of the domestic animals

TABLE II

Ratio	1875	1976
Cattle : man	0.79	0.084
Horse : man	0.081	0.0076
Sheep : man	1.49	0.28
Goat : man	0.38	0.0082
Tame reindeer : man	0.064	—

was low, as is evident from the animal : man ratio in the upper part of the Telemark region in 1875 as compared with 1976 (Table II).

In a recent study of the development of the population in south-western Norway,[22] it was stressed that, before about 1850, the production of the grain fields remained low and the need to feed the increasing population led to a marked expansion in field areas. It was only in the last half of the nineteenth century that new agricultural tools and ploughs and better fertilizing techniques increased the production of food and fodder and thus the milk and meat production of the cattle also increased. At the same time, emigration to America gained momentum. Milk and meat products from areas outside the farm during summer grazing were the most important parts of the farming. The home farms mainly served to produce fodder for the winter time, so that the cattle should not die from starvation. There was next to no production of milk in the winter season.

We have no early statistics as to the volume of the standing stock of forests in the nineteenth century, but we know that it was low. In 1903, Helland spoke in the following terms of the forests in Vest-Adger County, precisely the area where acidification of the lakes and rivers is serious

It will not be long before the forests in the inner parts of south-west Norway will be insufficient even to provide wood material for the coastal districts. The trees may renew themselves after having been logged, but because of the way in which logging takes place, the reforestation is often insufficient to form dense woods and heather, which needs a great deal of light, spreads among the sparse trees. The spreading of the heather prevents the tree seeds from germinating or even from coming in contact with soil, and the wood area thus steadily becomes more open whilst heather to the same degree spreads and increases in growth activity. The heather not only chokes the forest growth but also prevents grass from germinating. In order to counteract this evil, it was decided to burn off the vegetation which was considered harmful. It was found that the grass was renewed in the first years after burning the heathland. However, after some time, when the ash components had been leached, and the soil had further decreased in fertility, the heather spread again. Each time, the burning was followed by the growth of some young trees, since it made the areas open for tree seeds and for germination. The young wood would in many cases, if it had been left to develop, have spread, and we would again have got dense forests. Thus, it would have been possible by forest instead of fire to stop the spreading of the heather, which does not survive in the shade. The natural pace was too slow, however, and it was found easier to burn the whole plant vegetation once again; thus the total vegetation and the topsoil was burned and eroded. In this way the large heathland developed either completely barren or with only a few crippled pinewood trees.

Professor Helland's account has been confirmed by modern pollen analysis.[23]

Grazing outside farms was also followed by the introduction of trout and other fish into the rivers and lakes, which had had no fish population since the Ice age because waterfalls made natural invasion impossible. In great parts of southern Norway, the fish fauna, like the flora, have been dominated by the activity of man for more than 2500 years.

Although there are no statistics on the standing wood volume in the nineteenth century, the taxation data from 1927 to 1973 in the three southernmost counties of Norway are illustrative (Table IV).

The increase in the population and in the number of marginal farm units is illustrated by the Norwegian Statistical Central Bureau for a typical marginal farm area in southern Norway (Table III). (In south-western Sweden the same development is well documented.)

It is very difficult to assess the influence of biological change on the geochemistry in a given area. Data from one area may not be applicable in another. The investigations at Hubbard Brook[22] deal with hardwood ecosystems in New Hampshire at 44°N. Although the annual temperature in this area is not so very far from the temperatures in southern Scandinavia (60°N) the vegetation is different. Nevertheless, the measurements from Hubbard Brook show the same trend as the data found in the coniferous-forest ecosystems of Scandinavia and even in the impediment areas of southern Norway.

The accumulation of electropositive ions in trees is an important part of the geochemical cycle. Trees are long lived and the change from grazing land to reforested areas take a long time in Scandinavia; more than 60 years is necessary to grow a tree to such a size as to be worth logging. During this time, base accumulation is occurring in the standing stock of wood, and this process must be considered, together with the removal of bases from the area by timber export or in solution in the outflowing streams, in discussing the ion transport. In addition to this extraction of bases, the various types of vegetation produce "acids" in addition to those protons which are necessary

TABLE III

Year	Population	Number of farm units
1801	810	164
1865	1103	198
1900	805	184
1930	554	142
1950	394	123
1960	237	83
1973	106	46

TABLE IV

Area	Taxation year	Type of tree	10^6 m^3	%	Taxation year	Type of tree	10^6 m^3	%	Taxation year	Type of trees	10^6 m^3	%
		Spruce	13.35	49		Spruce	19.89	53		Spruce	21.49	52
		Pines	9.57	35		Pines	12.76	34		Pines	14.07	34
Telemark	1927	Leafwood	4.22	16	1954	Leafwood	5.03	13	1964/73	Leafwood	5.85	14
		Total	27.14	100		Total	37.68	100		Total	41.41	100
		Spruce	3.24	25		Spruce	6.27	30		Spruce	7.70	31
		Pines	6.79	53		Pines	10.26	49		Pines	11.97	48
Aust-Agder	1928	Leafwood	2.89	22		Leafwood	4.36	21	1964/73	Leafwood	5.34	21
		Total	12.92	100		Total	20.89	100		Total	25.01	100
		Spruce	0.14	2		Spruce	0.64	6		Spruce	1.20	9
		Pines	3.90	61		Pines	6.07	59		Pines	7.39	55
Vest-Agder	1928	Leafwood	2.39	37	1955	Leafwood	3.59	36	1964/73[a]	Leafwood	4.88	36
		Total	6.43	100		Total	10.30	100		Total	13.47	100
		Spruce	16.73	36		Spruce	26.80	39		Spruce	30.39	38
		Pines	20.26	44		Pines	29.09	42		Pines	33.43	42
Total	1927/28	Leafwood	9.50	20	1954/55	Leafwood	12.98	19	1964/73	Leafwood	16.07	20
		Total	46.49	100		Total	68.87	100		Total	79.89	100

[a] Standing steamwood in million cubic meter below bark based on *Landskogs Takseringens Beretninger*. Only trees with a stem diameter of more than 6 cm at a height of 1.5 m above ground level were considered.

in order to liberate ions from the mineral phace during the weathering processes. In humid–temperate to subarctic climates the formation of raw humus of very low pH is known to take place in reforested areas that were previously used as farming land.

The time necessary to develop a new podzol profile is of the order of 100 years. It is therefore not surprising that very delayed actions may be observed in studying the influence of changes in vegetation on the water chemistry of the stream. Although it was found in the investigations in the Hubbard Brook area that the stream water chemistry of an undisturbed forest ecosystem is highly predictable,[25] this is not the case in an area where there has been marked change from one type of culture plant (grass, green vegetables, potatoes, etc.) to another (spruce and pine). Nevertheless, in Norway excatly the same trend has been found as in the American investigations. In a catchment area the removal of electropositive ions (Ca, Mg, K, Na) in the outflowing streams and by accumulation in the growing biomass is larger than would correspond to the decrease due to protons from the precipitation. This demonstrates that additional proton sources must exist and even predominate over the protons brought in by precipitation and dry deposition.

The raw humus in a reforested coniferous area with an undergrowth of heather and blueberries can be illustrated by the following measurements. An area north of Oslo, where the land is known to have been cleared and used for grazing up to the last part of the nineteenth century, is now covered with spruce forests. Samples of the top—O and A—layers from this reforested area were collected and percolated by artificial "rainwater." Water samples of two different compositions were prepared. Both contained 5 mg Na_2SO_4 per liter; this is an amount of sulfur slightly higher than that in the rainwater of the most polluted areas in southern Norway. In addition, one solution contained 2.1 mg NaCl per liter—corresponding to a chlorine content typical of the more inland precipitation of Norway—and the other contained 21 mg NaCl per liter—corresponding to coastal rainfall. Both solutions were brought up to pH 8 by the addition of NaOH. The percolation was carried on in stages, each one corresponding to an average of 5 years of precipitation in southern Norway (about 7.5 years inland and 2.5 years in the south coastal area). The run-through time for each batch of "artificial rain" varied from one to a few hours. The observed pH in the percolate is shown in Figure 2.

This illustrates that the concentration of anions in the percolating solution is important in establishing the pH of the percolate. In both cases the pH of the percolate became very acid and even 100 years of precipitation did not bring the pH of the percolate up to 5.0. The irregularities of the individual results seem to be related to the rate of percolation and to the fact that

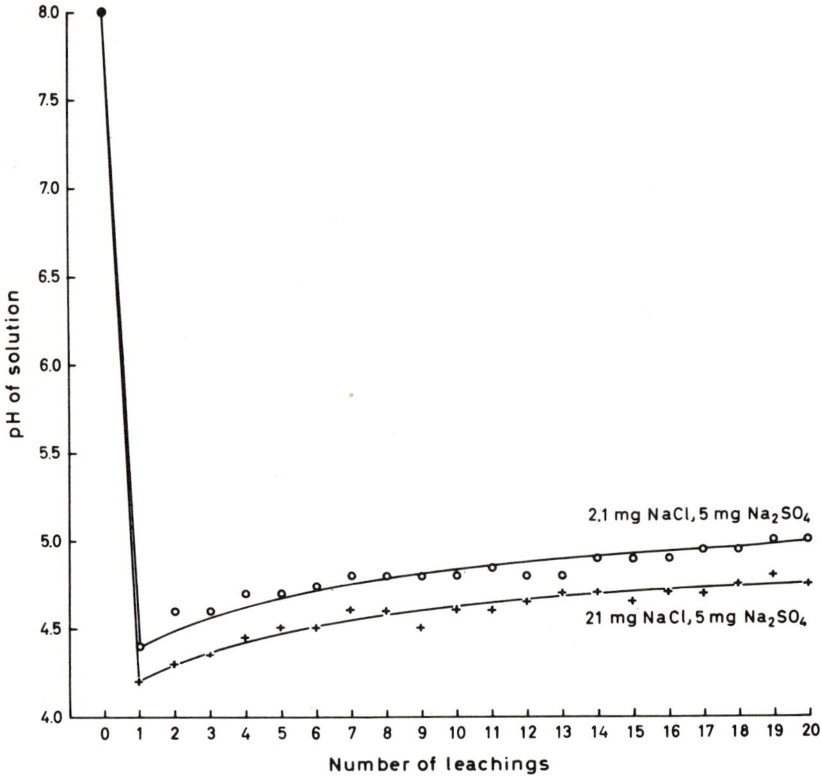

Figure 2 pH of artificial "rainwater" after percolation through raw humus from a reforested area. Initially, for each leaching, pH = 8. Each charge corresponds to approximately five years of precipitation in southern Norway. Even after a hundred years the humus layer will change alkaline input water into acid runoff water.

samples had to be left over night, so that there was a certain "recovery" of exchange capacity. Thus, a forest area which has developed within the last 90 years has an acid-exchange capacity from the litterfall and the raw humus which exceeds 100 years at present levels of precipitation, although obviously most of the acids produced in this forest area had been consumed or leached away during those 90 years. It has been shown[5] that topsoil from grassy areas on similar bedrock shows nearly neutral values for the pH even after the first percolation. "Rain" with 2.1 mg NaCl and 5 mg Na_2SO_4 gave a pH of 6.3 for the first filtrate and pH 7.1 for the second.

It is therefore easy to demonstrate the influence of the type of vegetation on the soil acidity, and it is equally easy to demonstrate that there is no watershed at which the pH is a direct result of the pH of the precipitation. In all the cases examined, the pH of the runoff water is a function of the intensity of the precipitation and possibly also of the ion content of the precipitation.

The intensity of the precipitation determines the flow path of the water in the ecosystem. For intense rain, a great part of the water is only percolating or running on top of the acid raw-humus layer, whereas for more moderate precipitation most of the water will pass through the weathering mineral soil underground, where the acids from the top soil are neutralized. Gneiss–granitic areas with sparse soil cover on top of the bedrock will normally be found to have very low conductivities in the river waters, and at the same time low buffering capacities. In such cases even a low percentage of acid water will have a large influence on the final pH value. In totally unbuffered systems, mixing 1 liter of water of pH 4 with 10 liters of water of pH 7 gives 11 liters of water of pH 5.

The S.N.S.F. research mentioned earlier,[2] the main objective of which was "to provide material for negotiations in order to limit the emission of SO_2 in Europe," found that every heavy episode of acid precipitation was followed by a wave of acid water in the streams running out of the catchment area. The conclusion drawn was that acids in the precipitation are the reason for the acidity in the river waters.

In order to investigate the truth of this statement is must be remembered that observations and experiments can only illustrate a theory; they can never prove it. What science can do is to try and disprove the hypothesis. The appropriate procedure in this case is to examine what happens to the pH in the stream when there is heavy rainfall which is cleaner than normal. Such investigations reveal that the pH drops in the same way as when the rainfall is very acid. In every catchment area studied it has been found that the pH is determined by reactions in the ecosystem and by the total ionic composition of the precipitation. Whether this precipitation contains acids or neutral salts does not make a great difference. In Norway, the acid part of the precipitation will, on average, lower the pH in streamwater by amounts that are insignificant in comparison with the annual variations.

As the annual natural production of acids in most reforested areas, moss or heather lands is found to be 5–10 times higher than the amount of acids in the annual rain of south Norway, it may be concluded that, if the rain had been neutral rather than acid, the pH of the rivers draining the area would, on average, have been higher by ~0.04–0.08. These figures are small in comparison with the variations in pH over the year.

It is always the intensity of the rain rather than its acidity which determines the pH of the rivers and creeks in southern Norway; this is illustrated by Figure 3, which is based on data from the Birkenes area in southern Norway.†

† The data have been kindly put at my disposal by the Norwegian Acid Precipitation Project.

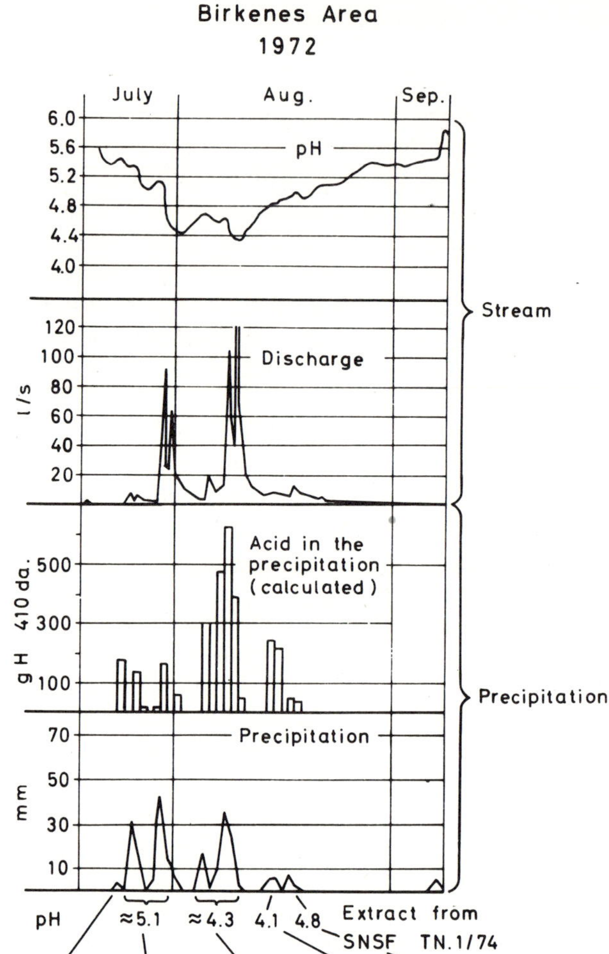

Period :	24/7	26/7–31/7	5/8–9/8	14/8–15/8	17/8
Precipitation :	5.6 mm	107 mm	86 mm	11.5 mm	7.6 mm
pH :	4.1	4.8–5.55	3.9–4.4	3.9–4.1	4.75

Figure 3 pH in the precipitation and runoff water in the Birkenes area of southern Norway (summer, 1972).

After a dry summer period in 1972 with a relatively high pH in the creek, there was a heavy fall (106 mm) of mainly unpolluted rain in this catchment area at the end of July. As a result, the pH of the creek dropped from 5.6 to 4.4 (i.e., 5 times the acidity of the rainwater, from $10^{-5.1}$ mol.H^+ per liter to $10^{-4.4}$ mol. per liter). As the pH started to recover in August, further heavy fall (86 mm), this time of very acid rain, pH 3.9–4.4, brought the pH in the creek down to the same level as in late July. This time the river water was less acid than the rainwater.

Similar experiments carried out by the Norwegian Acid Precipitation Group illustrate this point.[2] In two parallel series, the following solutions were allowed to percolate through raw humus of natural pH 4.2:

0.01 meq H_2SO_4 per liter, i.e., pH 5;
0.1 meq H_2SO_4 per liter, i.e., pH 4;
1 meq H_2SO_4 per liter, i.e., pH 3.

In these series the solution with 1 meq H_2SO_4 per liter gave a filtrate of pH 3.9 in one case, and a filtrate of pH 3.6 in the other, while the solution with 0.01 meq H_2SO_4 per liter gave a filtrate of pH 4.15 in one case and a filtrate of pH 5.2 in the other. The solution with 0.1 meq sulfuric acid per liter gave intermediate values.

Obviously the mineral components in the humus soil had a certain action in neutralizing some of the sulfuric acid. It might have been concluded from these experiments that the pH of the original solution had a considerable influence on the pH of the runoff water.

At the same time, however, the Acid Precipitation Project also carried out experiments with neutral NaCl solutions in the same proportions[2]:

0.01 meq NaCl per liter;
0.1 meq NaCl per liter;
1 meq NaCl per liter.

In these series the solution with 1 meq NaCl per liter gave a filtrate of pH 3.8 in one case, and a filtrate of pH 3.8 in the other, while the solution with 0.01 meq NaCl per liter gave a filtrate of pH 4.2 in one case and a filtrate of pH 4.9 in the other. Again, the solution with 0.1 meq NaCl per liter gave intermediate values.

As will be seen, the neutral solution of NaCl ended up with considerably lower pH than the corresponding solution of H_2SO_4; the pH of the solution with 1 meq NaCl per liter remained constant at 3.8. Thus it is quite clear that what is important is not the pH of the initial solution, but the total ionic strength. And it is further demonstrated that the pH of the percolate depends on the type of raw humus or the ion-exchange system as a whole.

According to the S.N.S.F. report,[2] there has been a drop of 1–2 in pH in some Norwegian watercourses during the last 30–40 years. This means that the acidity has gone up by a factor of between 10 and 100. At the same time it is thought that the sulfate content in the atmosphere of north-western Europe has probably increased by a factor of two, as a result of the increased combustion of sulfur-bearing fuel. It is now clear that it is the total content of electrolyte in the precipitation which determines the pH of the runoff water from any given catchment area. Thus the maximal increase in the electrolyte content would have been a doubling, if no other electrolytes had been present. This would mean a decrease in the pH of 0.3 (i.e., log 2) but, because the chlorine content and natural sulfates have not been influenced by the increased combustion of sulfur-bearing fuel the relative concentration of total electrolytes cannot have been doubled. The decrease in pH due to the increase of electrolyte must therefore have been considerably less than 0.3. Other conditions in the catchment area being equal, a 50% increase in the total electrolyte content would involve a decrease in pH by 0.18 and a 25% increase would involve a decrease in pH by 0.1.

It must be remembered that the pH values in episodes of acid rainfall in Scandinavia correspond to the pH values which would result if a thin slice of lemon were squeezed into 50 liters of distilled water, and the average pH in the acid regions corresponds to one slice of lemon in 200 liters of distilled water. Laymen often say: "But it is obvious that, if the input water becomes more acid, the output must also become more acid. Isn't that so?"

To this it must be said that the water is passing through a complicated ecosystem from the moment it hits the ground until the time it reaches the watercourses. Because of the low acidity in the rainwater, the above statement could as well have been: "But it is obvious that if a person is changing his diet, so that he is drinking one extra cup of tea with a slice of lemon in it each month, his urine must become more acid. Isn't that so?".

REFERENCES

1. P. J. Dillon *et al.*, "Acid lakes in Ontario, Canada. Their extent and response to base and nutrient addition," in *Jubilee Symposium on Lake Metabolism and Lake Management* (Uppsala, 1977).
2. S.N.S.F., *Impact of Acid Precipitation on Forest and Freshwater Ecosystem in Norway*, S.N.S.F. Project Research Report No. 6 (Ås, NLH, Norway, 1976).
3. T. C. Hutchinson and L. M. Whitby, *The Effect of Acid Rainfall and Heavy Metal Particulates on a Boreal Forest Ecosystem near Sudbury Smelting Region of Canada* (1972).
4. O.E.C.D., *The O.E.C.D. Programme on Long-range Transport of Acid Pollutants. Measurements and Findings* (O.E.C.D., Paris, 1977).
5. I. Th. Rosenqvist, *Sur Jord—Surt Vann* (Acid Soil—Acid Water) (Ingeniørforlaget A/S, Oslo, 1977).
6. A. Helland, *Lister og Mandals Amt* (Norges Land og Folk, Kristiania, 1903).
7. R. C. Reynolds and N. M. Johnson, "Chemical weathering in the temperate glacial environment of the Northern Cascade Mountains," *Geochim. Cosmochim. Acta*, **36**, 537–559 (1972).
8. H. Rueslåtten, *En Kvartærgeologisk Kartlegging av Dagali-området med Mineralogisk Undersøkelse av Podzolforvitring i Moreneavsetningen*, Thesis (M.-Nat. Fak., Univ. Oslo, 1976).
9. I. Jensen, I., *Kjemisk Forvitring i Prekambrisk, Kambrosiluriske og Permiske Bergarter i Numedal. En Limnologisk Undersøkelse i 1970–71*, Thesis in Limnology (Univ. Oslo, 1972).

10. H. Sioli, "Studies in Amazonian waters," in *Atas to Simposio Sobre a Diota Amazonia*, Vol. 3, *Limnologia* (1977) pp. 9–50.
11. H. Klinge and W. Ohle, "Chemical properties of rivers in the Amazonian area in relation to soil conditions," *Verh. Internat. Verein Limnol.*, **15**, 1067–1076 (1961).
12. K. M. Strøm, "pH values in Norwegian mountain lakes," *Nytt Magaz. F. Nature*, **IXII** (1925).
13. R. S. Clymo, "The origin of acidity in *Sphagnum* bogs," *The Biologist*, **67**, 427–430 (1967).
14. R. S. Clymo, "The origin of acidity in *Sphagnum* bogs," *Proceedings of a Symposium* (Amsterdam–Nienversliens, 1967) 273–284.
15. H. Bergseth, "Relation zwischen aciditert und wegetationstyp norwegischer waltböden," *Act. Agricultura Scandinavica*, **27** (1977).
16. L. Wiklander, "The acidification of soil by acid precipitation," *Grunnförbettring*, **26**, 155–164 (1974).
17. M. Häberle, "Luftverschmutzung": weiss der Mensch was er und die Natur tut? *Verein Industrieller Kraftwerkbetriebe*, 94–104 (1973).
18. C. Bonnevie-Svendsen and O. Gjems, "Amount and chemical composition of the litter from larch, beech, Norway spruce and Scots pine stands and its effect on the soil," *Medd. Det Norske Skogførsoksvesen*, No. 98 (1957).
19. L. E. Wise (editor), *Wood Chemistry* (Reinhold, New York, 1946).
20. L. Wiklander, "Forsuringsproblemet för mark och vatten. II," *Skogs- och Lantbruk Akad. Tidskr.*, **116**, 93–108 (1977).
21. B. Haugen, *Fra Seter til Gård* (Norsk Arkeologisk Selskap, 1947).
22. E. S. Pedersen, *Befolkningsutviklingen på 1800-tallet fra Haug O.K.*, HEIDNI Tidsskrift fra Rogalands Arkeologiske Forening No. 4 (1977).
23. A. Skogen, *Den Vest-norske Lyngheien—et Kulturlandskap i Endring*, Forskningsnytt No. 4 (1974).

I. Th. Rosenqvist
University of Oslo

OXIDATION AND OXIDATION–INHIBITION OF SULFUR DIOXIDE

Transformations of sulfur dioxide in the atmosphere to sulfite and sulfate particles and to solutions of the same ions carry large-scale and long-range health and ecological implications (the health effects of submicron aerosols, acid rain). Already in *Advances in Environmental Science and Engineering*, Ziegler has discussed the important mechanisms of oxidation in the atmosphere and one of the major effects which—at least in part—results from such transformations: acid rain.[1,2] The purpose of this article is to review additional evidence bearing on the oxidation mechanisms, particularly the inhibition of that oxidation and its potential relevance to the atmosphere.

Absorption into water (in the form of rain and cloud drops) may be one of the major routes for the removal of sulfur dioxide from the atmosphere. Absorbed sulfur dioxide forms hydrosulfite and sulfite; the relative amounts are governed by the pH of the solution and the temperature. Sulfite ion can be oxidized to sulfate, and this reaction is catalyzed by transition metal salts[3–11] and reagents such as ozone.[12,13] General mechanisms for these reactions have been known for a long time,[3–5,8,9,14] although certain details remain unknown or are poorly quantified. Sulfur dioxide in plumes can apparently be oxidized by both heterogeneous and homogeneous pathways.[15,16]

Results on the uncatalyzed oxidation of atmospheric sulfur dioxide by oxygen in aqueous systems should be examined first; these have recently been summarized by Beilke *et al.*[17] It is perhaps unnecessary to point out that true "lack of catalysis" is not substantiated in all of these data. The rate of sulfate formation, $d[SO_4^{2-}]/dt$, was given as the product of a normalized rate Q and the sulfur dioxide partial pressure,

$$\frac{d[SO_4]}{dt} = QP_{SO_2},$$

and Q was expressed in terms of equilibrium constants for the SO_2–sulfite system and k_0, the overall first-order rate constant for sulfite-ion oxidation; Q was found to be proportional to $[H^+]^{-2}$. The value of k_0 is not known with high precision and is usually taken from the work of one or more previous investigators.[17–20]

The catalyzed reaction has been written as a free-radical chain reaction or a complexing mechanism, or a combination of both. The sequence below (or modifications thereof) is usually considered.

Initiation

$$SO_3^{2-} + M^{+i} \xrightarrow{k_1} M^{+i-1} + SO_3^{\cdot -}, \quad (1a)$$

$$HSO_3^- + M^{+i} \xrightarrow{k_{1a}} M^{+i-1} + SO_3^{\cdot -} + H^+. \quad (1b)$$

Propagation

$$SO_3^{\cdot -} + O_2 \xrightarrow{k_2} SO_5^{\cdot -}, \quad (2)$$

$$SO_5^{\cdot -} + SO_3^{2-} \xrightarrow{k_3} SO_5^{2-} + SO_3^{\cdot -}. \quad (3)$$

Termination

$$\left.\begin{aligned} 2SO_5^{\cdot -} &\xrightarrow{k_4} (S_2O_{10}^{2-}) \xrightarrow{k_{4a}} 2SO_4^{2-} + O_2, \\ 2SO_3^{\cdot -} &\xrightarrow{k_{4b}} S_2O_6^{2-}. \end{aligned}\right\} \quad (4)$$

Disproportionation

$$SO_5^{2-} + SO_3^{2-} \xrightarrow{k_5} 2SO_4^{2-}. \quad (5)$$

The formation of higher oxyacids of sulfur is also possible. A general rate expression for this mechanism takes the form

$$\frac{d[SO_4^{2-}]}{dt} = k[M^{+i}]^x[O_2]^y[SO_3^{2-}]^z[H^+]^{-n},$$

where k is a composite rate constant and n, x, y and z can range from zero to 2; i denotes the oxidation state of the metal. In part, the range in these values is due to the different conditions under which the reaction has been studied. These conditions have included the completely homogeneous system (oxygen and sulfite are in separate aqueous solutions which are mixed and the catalyst is in either the oxygen or the sulfite solution) and the heterogeneous system (oxygen mass transfer into sulfite solution containing catalyst as well as oxygen and sulfur-dioxide mass transfer into an aqueous solution of the catalyst).[7,8,21–23] Although the counter ion in these oxidations is generally ignored, it should be pointed out that the relative rates of the cobalt-catalyzed oxidation of sodium, potassium and ammonium sulfite are 5 : 3 : 1 in the homogeneous system.[24]

The complexing mechanisms—using Co^{2+}—have been written as[23]

$$Co^{2+} + SO_3^{2-} \xrightleftharpoons{\text{fast}} \text{complex},$$

$$\text{complex} + O_2 \xrightarrow{\text{slow}} SO_5^{2-} + Co^{2+},$$

$$SO_5^{2-} + SO_3^{2-} \xrightarrow{\text{fast}} 2SO_4^{2-},$$

and for iron[9]

$$Fe(OH)^{2+} + H^+ \rightleftharpoons Fe^{3+} + H_2O,$$

$$Fe^{3+} + HSO_3^- \rightleftharpoons Fe(HSO_3)^{2+},$$

$$Fe(HSO_3)^{2+} + SO_3^{2-} \rightleftharpoons Fe(HSO_3)(SO_3),$$

$$Fe(HSO_3)(SO_3) + O_2 + H_2O \longrightarrow Fe(OH)^{2+} + 2HSO_4^-,$$

$$Fe(HSO_3)(SO_3) + Fe^{3+} + H_2O \rightleftharpoons 2Fe^{2+} + HSO_4^- + H^+HSO_3^-.$$

The existence of $[Mn(SO_3^{2-})_3]^{4+}$ complexes has been suggested.[6] Schmidkunz[25] proposed a $^{\cdot}O_2H$ chain

$$^{\cdot}O_2H + SO_3^{2-} \longrightarrow SO_4^{2-} + {}^{\cdot}OH,$$

$$^{\cdot}OH + SO_3^{2-} \longrightarrow SO_4^{2-} + H^{\cdot},$$

$$^{\cdot}H + O_2 \longrightarrow {}^{\cdot}O_2H.$$

Note that $^{\cdot}OH$ could also arise from the reductions of metal ions such as Co^{3+}.[26] Also, the step

$$Fe^{3+} + HSO_3^- \longrightarrow Fe(HSO_3)^{2+}$$

is formally equivalent to

$$\longrightarrow Fe^{2+} + HSO_3^{\cdot}.$$

Thus, possible chain carriers for the reaction are $SO_3^{\cdot-}$, $SO_5^{\cdot-}$, $HSO_3^{\cdot}$, $OH^{\cdot}$ and $O_2H^{\cdot}$, and these may even be involved in the mechanisms written as complexing steps. Electronic spectra for several of the sulfur-radical ions have been reported.[14] Electron-spin-resonance studies, the presence of induction periods and the action of typical antioxidants (see below) support the formulation of the reaction as a free-radical chain.[3–5,26–28]

A modification of the Backstrom mechanism involving free radicals has also been proposed to explain ozone oxidation.[12]

The mechanism of the inhibition has to be considered in relation to these oxidation mechanisms. References on the inhibitions of sulfite oxidation date back to the last century[4]; the most recent general summary of this work is given by Schroeter.[5] While there has been extensive work on inhibition and it has been found that a wide variety of substances appear to inhibit the reaction by varying degrees, there is little quantitative agreement and even some contradiction in the literature. Cherkov[29] found *p*-phenylene diamine to be an effective inhibitor in solutions not "contaminated" by catalytic solids; the efficiency of the catalyst was reportedly decreased by a factor of 3–4 in the presence of fly ash from flue gas; fly ash may contain phenolic inhibitors, i.e., it may provide both the inhibitor and the catalyst for the oxidation. Although there is some evidence on this, detailed investigations are lacking.[30]

Organic compounds in rather dilute solutions (10^{-6}–10^{-4} M) representing the following wide variety of classes are known to inhibit sulfite oxidation: acids, alcohols, glycols, mono- and polysaccharides, amines, amides, aldehydes, alkaloids, ketones, indoles, phenols and quinones.[5,31,32] The following order of effectiveness for organic inhibitors was reported: hydroquinone > pyrogallol > 2-naphthol > 1-naphthol > pyrocatechol > resorcinol > phenol > phloroglucinol. Although there is little evidence for synergistic effects (see below), such effects can certainly be envisioned for such diverse groups of compounds.

Fuller and Christ[33] and Backstrom[3] found mannitol to be an effective inhibitor and noted that the inhibitory effect of mannitol on sulfite oxidation was uniform for mannitol concentrations in the range 10^{-7}–10^{-2} M. Inorganic anions such as arsenite, antimonite, phosphite and cyanide may also function as chain terminators.

Little quantitative information on the effect of inhibitors on the kinetics of the oxidation has been reported; Schroeter has described the inhibitor effect by the equation[5,34]

$$-\frac{dS_t}{dt} = \frac{k_0 S_t A}{B + m},$$

where $S_t = [HSO_3^-] + [SO_3^{2-}]$; k_0 is the specific rate constant of the uninhibited, uncatalyzed reaction ($\sim 2.5 \times 10^{-3}\ s^{-1}$); A and B are constants characteristic of the inhibitor and rate of oxygen absorption; m is the molar concentration of inhibitor.

The effectiveness of compounds such as phenols in inhibiting the oxidation is in general agreement with a radical-chain mechanism.[35] This aspect can be discussed together with the mechanism.

For Eqs. (1a), (2)–(4) the following expressions may be written for the rates

$$r_2 = \frac{-d[O_2]}{dt} = -\frac{d[SO_3^{\bar{\cdot}}]}{dt} = k_2[SO_3^{\bar{\cdot}}][O_2],$$

$$r_3 = k_3[SO_5^{\bar{\cdot}}][SO_3^{2-}],$$

where the subscript corresponds to the equation number to which the rate applies.

Suppose that $r_2 > r_3$ (i.e., zero order in oxygen), that

$$[SO_3^{\overline{\cdot}}] < [SO_5^{\overline{\cdot}}]$$

that the rate of termination

$$r_4 = k_4[SO_5^{\overline{\cdot}}]^2$$

and

$$[SO_5^{\overline{\cdot}}] = r_4^{1/2}k_4^{-1/2};$$

then

$$\frac{-d[O_2]}{dt} = r_4^{1/2}\frac{k_3[SO_3^{2-}]}{k_4^{1/2}} = \frac{d[SO_4^{2-}]}{dt} = r_u,$$

where r_u is the uninhibited rate.

Presumably, $r_1 \propto [M^{+i}]^x$; various values of x have been reported in the literature. For $r_3 > r_2 = -k_2[SO_3^{\overline{\cdot}}]$, $[O_2]$ is first order with respect to oxygen. (This situation arises in hydrocarbon oxidation at oxygen partial pressures below 200 mm Hg; this point is made here since the sulfite oxidation shows formal similarities to hydrocarbon oxidation.)

For the inhibited case, the following equations can be added

$$SO_5^{\overline{\cdot}} + HIN \longrightarrow HSO_5^- + IN^{\cdot}, \tag{6}$$

$$SO_3^{\overline{\cdot}} + HIN \longrightarrow HSO_3^- + IN^{\cdot}. \tag{7}$$

Assuming that $k_6 > k_7$, r_a, the rate of oxidation in the presence of the inhibitor (HIN), is

$$\begin{aligned} \frac{-d[SO_5^{\overline{\cdot}}]}{dt} &= k_6[SO_5^{\overline{\cdot}}][HIN] \\ &= k_6 r_4^{1/2}k_4^{-1/2}[HIN], \\ r_a &= r_u^2 k_4 k_3^{-1} k_6^{-1}[HIN]^{-1}[SO_3^{2-}]^{-1}, \end{aligned}$$

or

$$\frac{k_4}{k_3 k_6} = \frac{r_a}{r_u^2}[HIN][SO_3^{2-}] = k. \tag{8}$$

Since k_4 and k_3 are functions of sulfite only, this relationship offers a comparison of different antioxidants at different concentrations, or $1/k$ is a quantitative measure of the chain-breaking efficiency of the antioxidant.

When HIN is a phenolic compound the reaction can be formulated as follows

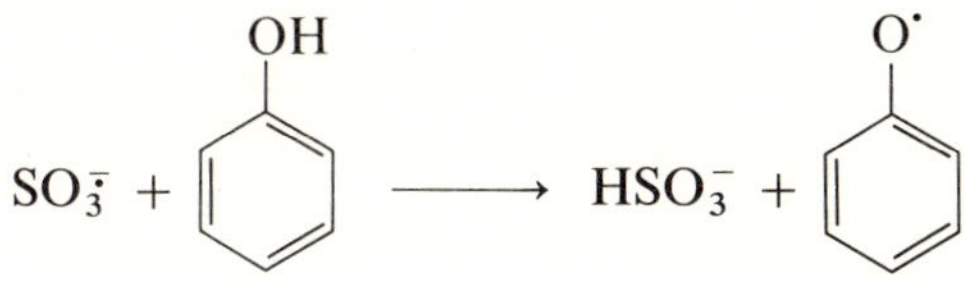

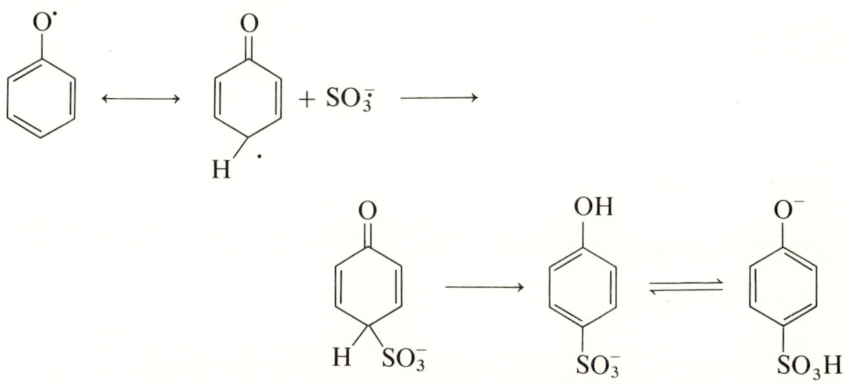

or $C_6H_5OSO_3H$. This sequence offers the potential of generating a new inhibitor, and thus reducing the actual amount of HIN needed for effective inhibition. Some evidence on this point is available; phenol sulfonic acid has some inhibitor properties.[32]

Equation (8) can be condensed to the following form

$$\frac{r_a}{r_u^2}[\mathrm{HIN}]^n = k'. \tag{9}$$

We have investigated this relationship with different types of antioxidants.[36] Figure 1 shows the results obtained in a cobalt-catalyzed system with hydroquinone ($n = 1.5$) and *p*-phenylene diamine ($n = 1$); 4-methoxyphenol and resorcinol were found to be far less efficient. Several other compounds—including ascorbic acid—showed antioxidant properties.

Hayon *et al.* replaced reaction (3) by

$$SO_5^{\overline{\cdot}} + SO_3^{2-} \longrightarrow SO_4^{2-} + SO_4^{\overline{\cdot}}$$

$$SO_4^{\overline{\cdot}} + SO_3^{2-} \longrightarrow SO_4^{2-} + SO_3^{\overline{\cdot}}$$

on the basis of their work on the inhibitory effect of alcohols[14]; i.e., $SO_4^{\overline{\cdot}}$ is scavenged more effectively than $SO_3^{\overline{\cdot}}$ or $SO_5^{\overline{\cdot}}$. If this is the correct sequence there should be no induction period for sulfate formation. Schmidt suggested that hydroquinone functioned in competition with sulfite for oxygen.[37]

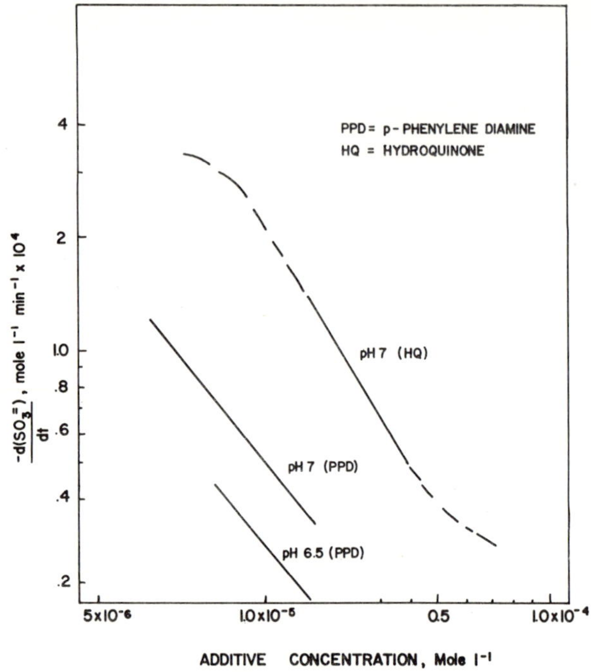

Figure 1 Effect of additive on sulfite oxidation rate.

In general, antioxidants show a lengthening of the induction period, a decrease in the rate of oxidation or both.[5,32,36] The intensity of light emission is also decreased.[27]

Applications of these findings to atmospheric processes must examine both the chemistry and the mass-transfer aspects of the problem. The principal interest centers on transformations in droplets of various sizes (fog, cloud, rain). These drops can capture particles containing sulfites and sulfates or these particles may serve as condensation nuclei, in which case initial sulfur-species concentrations could be very high; sulfur dioxide could also adsorb on particles.[38] If these salts are in soluble form, the equilibria involving HSO_3^-, SO_3^{2-}, HSO_4^- and SO_4^{2-} are usually assumed to be established instantaneously.[39] The oxidation rate is affected by catalytic ions which are either already present or are likewise captured as aerosols. Of the catalytic ions Fe^{2+}, Mn^{2+} and Co^{2+}, only Fe^{2+} appears to be present in the atmosphere in reasonable amounts,[6,40,41] although estimates of high manganese concentrations have been reported.[10] Ranges for dissolved metal ions in rainwater in various parts of the world are (ppb): Fe, 10–230; Mn, 0.19–25; and V, <4.7.[42] In the oxide form (Fe_2O_3, V_2O_5) catalytic properties are absent but conversion to active species can take place.

Ambient measurements have been made of organic compounds that could act as potential inhibitors. These range from 1 ng/m^3 in clean air[43] to 10 μg/m^3 for individual phenolic compounds,[44] i.e., 30 μg/m^3 for phenol and 0.02 μg/m^3 for 4-hydroxybenzaldehyde. If an average of 1 μg/m^3 of water-soluble inhibitor molecules of average molecular weight 110 were assumed to be in particle form, this could lead to an average concentration of $\sim 5 \times 10^{-5}$ M in cloud or rainwater, a value that is well within the range that could be effective (Figure 1).

Barrie[45] absorbed sulfur dioxide into droplets containing manganese chloride and phenol. Only when the phenol concentration was ten times the manganese-chloride concentration did the phenol slow the absorption rate of sulfur dioxide. Since this result was based on the analysis of sulfur as sulfate it probably represents the combined effect of absorption and oxidation. Nevertheless, after 30-min exposure of the drops to 480 ppb sulfur dioxide, drops with 10^{-5} M manganese chloride were found to contain approximately three times as much sulfur as drops with both manganese chloride (10^{-5} M) and phenol (10^{-4} M). In our batch experiments, which were performed at constant pH, the SO_2 absorption rate was not affected by the metal-ion concentration or the ratio of catalyst to inhibitor.[32,36]

Several unidentified phenolic compounds have been reported in particle samples from urban air; 2-fluorenol was positively identified in New York City air.[46]

Such compounds could come from numerous sources; many of these have been identified.[47–49] Quantitative information on emissions is less certain; this, in turn, makes estimates of ambient concentrations in plumes or during long-range transport uncertain. Phenolic compounds arise from the industrial processes manufacturing and using such compounds. They have also been reported from automobile exhaust, tobacco smoke, plastics incineration, petroleum refining, varnish cookers, beehive coke ovens, coke-conversion plants, phenol-resin plants, wire-coating ovens, Kraft black-liquor pyrolysis, meat smokehouses and wood smoke. Some specific identifications include phenols and nitrophenols from Diesel exhaust as well as benzaldehyde, guaiacol and 2,6-dimethoxypyrogallol from wood burning. Agricultural burning generates oxygenated organics, which may include phenols, aldehydes, and acids. Aliphatic and aromatic amines result from many industrial processes. Substituted *p*-phenylene diamines are used in tires as antiozonants; they may enter the atmosphere as a result of tire wear and volatilization. Aldehydes and other complex organics are also formed in the later stages of photochemical-smog formation; such compounds are potential inhibitors.

The importance of the oxidation–inhibition should be documented by the measurement of oxidation rates for sulfur dioxide and by the occurrence of S^{4+}- species in cloud and rainwater. Oxidation rates for sulfur dioxide in power-plant and urban plumes vary widely,[15,50] and it is conceivable that the observed net rates reflect the combined effects of catalysis–inhibition in addition to dilution. In this sense the transformation can be accommodated by the formalisms proposed recently[50]

$$SO_2 + P \longrightarrow SO_4^{2-},$$

where P can be though of as a limiting reagent (or combination of reagents) that is nonstoichiometric with respect to sulfur dioxide. In actuality, no measurements of trace metals and potential inhibitors in plumes are available to test this possibility. As far as cloud and rainwater are concerned, many measurements of sulfate were performed on samples that were not handled in a way specifically calculated to prevent sulfite oxidation. Some investigators have reported sulfite in rainwater,[51,52] and more reports can be anticipated in the future. Some recent ratios (based on concentrations in μM of sulfur dioxide to sulfate in rainwater are: 0.21,[53] 0.50[53] and 0.33.[54] The sulfite values used in these ratios were comparable to the value reported by Hogstrom, 0.29×10^{-5} M.[55] Generally, however, the ratios are much lower. Finally, it must be emphasized that the presence of sulfite *per se* does not prove that inhibition has occurred. Simultaneous analytical data on potential catalysts and inhibitors are needed.

In the analysis of these processes, consideration must also be given to the time frame and to the mass-transfer aspects. Basically, two time regimes are available: hours-days (fog and cloud drops, plumes) and seconds-minutes (raindrops). Trace-gas scavenging has been studied theoretically and experimentally by many authors, in both laboratory situations and large-scale field experiments.[6,56–58] The process can be modeled in detail for the first two of the three steps generally considered: gas-phase mass transport, dissolution and liquid-phase mass transport. Oxygen gas-phase mass transport is not likely to be limiting, except in very early stages of plume travel when p_{O_2} may be less than 0.1 atm.[36] The neutralizing effect of ammonia may be too small to affect washout significantly.[59]

An adequate assessment of this problem in more quantitative terms will require substantial ambient-data input and iterations on models for wet and dry removal of particles and gases.[60]

REFERENCES

1. E. N. Ziegler, "Sulfate-formation mechanism: Theoretical and laboratory studies," in *Advances in Environmental Science and Engineering*, Vol. 1, edited by J. R. Pfafflin and E. N. Ziegler (Gordon and Breach, New York, 1979).
2. E. N. Ziegler, "Precipitation composition: Northeastern United States," in *Advances in Environmental Science and Engineering*, Vol. 1, edited by J. R. Pfafflin and E. N. Ziegler (Gordon and Breach, New York, 1979).

3. H. L. J. Backstrom, "The chain reaction theory of negative catalysis," *J. Am. Chem. Soc.*, **49**, 1460 (1927).
4. S. L. Bigelow, "Katalytische Wirkungen auf die Geschwindigkeit der Oxidation des Natriumsulfits durch Sauerstoff," *Z. Phys. Chem.*, **26**, 493 (1898).
5. L. C. Schroeter, *Sulfur Dioxide* (Pergamon, Oxford, 1966).
6. L. A. Barrie and H. W. Georgii, "An experimental investigation of the absorption of sulfur dioxide by water drops containing heavy metal ions," *Atm. Env.*, **10**, 743 (1976).
7. D. R. Coughanowr and F. E. Krause, "The reaction of SO_2 and O_2 in aqueous solutions of manganese sulfate," *I. and E.C. Fundamentals*, **4**, 61 (1965).
8. J. E. Sawicki and C. H. Barron, "On the kinetics of sulfite oxidation in heterogeneous systems," *Chem. Eng. J.*, **5**, 153 (1973).
9. J. Freiberg, "The mechanism of iron-catalyzed oxidation of SO_2 in atmospheric aerosols," *Atm. Env.*, **9**, 661 (1975).
10. R. T. Cheng, M. Corn and J. O. Frohliger, "Contribution to the reaction kinetics of water soluble aerosols and SO_2 in air at ppm concentrations," *Atm. Env.*, **5**, 987 (1971).
11. P. Brimblecombe and D. J. Spedding, "The catalytic oxidation of micromolecular aqueous sulfur dioxide," *Atm. Env.*, **8**, 937 (1974).
12. T. V. Larson, N. R. Morike and H. Harrison, "Acid sulfate aerosols: Formation from heterogeneous oxidation by O_3 in clouds," *Atm. Env.* (1978).
13. R. A. Cox and S. A. Penkett, "Effect of relative humidity on the disappearance of ozone and sulfur dioxide in contained systems," *Atm. Env.*, **6**, 365 (1972).
14. E. Hayon, A. Treihin and J. Wilf, "Electronic spectra, photochemistry, and autooxidation mechanism of the sulfite–bisulfite–pyrosulfite systems. The SO_2^-, SO_3^-, SO_4^-, SO_5^- radicals," *J. Am. Chem. Soc.*, **94**, 47 (1972).
15. P. M. Foster, "The oxidation of sulfur dioxide in power station plumes," *Atm. Env.*, **3**, 157 (1969).
16. H. W. Sidebottom, C. C. Bradcock, G. E. Jackson and J. G. Calvert, "Photooxidation of sulfur dioxide," *Env. Sci. Techn.*, **6**, 72 (1972).
17. S. Beilke, D. Lamb and J. Müller (1975), "On the uncatalyzed oxidation of atmospheric SO_2 by oxygen in aqueous systems," *Atm. Env.*, **9**, 1083 (1975).
18. H. A. C. McKay, "The atmospheric oxidation of sulfur dioxide in water droplets in the presence of ammonia," *Atm. Env.*, **5**, 7 (1971).
19. W. D. Scott and P. V. Hobbs, "The formation of sulfate in water droplets," *J. Atm. Sci.*, **24**, 54 (1967).
20. J. M. Miller and R. de Pena, "Contribution of scavenged sulfur dioxide to the sulfate content of rainwater," *J. Geophys. Res.*, **77**, 5905 (1972).
21. T. Chen and C. H. Barron, "Some aspects of the homogeneous kinetics of sulfite oxidation," *I. and E.C. Fundamentals*, **11**, 466 (1972).
22. R. D. Srivastava *et al.*, "The kinetics of oxidation of sodium sulfite," *Can. J. Chem. Eng.*, **46**, 181 (1968).
23. P. V. Danckwerts, *Gas–Liquid Reactions* (McGraw–Hill, New York, 1970).
24. G. C. Mishra and R. D. Srivastava, "Homogeneous kinetics of potassium sulfite oxidation," *Chem. Eng. Sci.*, **31**, 969 (1976).
25. H. Schmidkunz, *Chemilumineszenz der Sulfitoxidation*, Ph.D. Thesis (University of Frankfurt, Frankfurt a.M., 1963).
26. W. A. Waters, "Roles of cobaltic salts as catalytsts of oxidation," *Disc. Faraday Soc.*, **46**, 158 (1968).
27. J. Stauff, "Chemiluminescence of some reactions with oxygen," *Photochem. Photobiol.*, **4**, 1199 (1965).
28. S. Bengtsson and I. Bjerle, "Catalytic oxidation of sulfite in dilute aqueous solution," *Chem. Eng. Sci.*, **30**, 1429 (1975).
29. B. A. Chertkov, "General equation for the oxidation of sulfite–bisulfite solutions in the extraction of SO_2 from gases," *J. Appl. Chem. U.S.S.R.*, **34**, 743 (1961).
30. H. F. Johnstone, "Metallic ions as catalysts for the removal of sulfur dioxide from boiler furnace gases," *Ind. Eng. Chem.*, **23**, 559 (1931).
31. B. Dev and B. D. Jain, *J. Sci. Ind. Res. India*, **20**, 461 (1961).

32. E. R. Altwicker and T. Sekulic, "Inhibition of sulfur dioxide oxidation during absorption," *Env. Lett.*, **7**, 125 (1974).
33. E. C. Fuller and R. H. Christ, "The rate of oxidation of sulfite ions by oxygen," *J. Am. Chem. Soc.*, **63**, 1644 (1941).
34. L. C. Schroeter, "Kinetics of air oxidation of sulfurous acid salts," *J. Pharm. Sci.*, **52**, 559 (1963).
35. L. R. Mahoney, "Antioxidants," *Angew. Chem. Internat. Edit.*, **8**, 547 (1969).
36. E. R. Altwicker, "Sulfur dioxide absorption, oxidation, and oxidation–inhibition," *Dechema Monographien*, **80**, No. 1639–69, Part 2, 343 (1976).
37. M. Schmidt, "Fundamental chemistry of sulfur dioxide removal and subsequent recovery *via* aqueous scrubbing," *Int. J. Sulfur Chem. B*, **7**, 11 (1972).
38. L. D. Hulett, T. A. Carlson, B. R. Fish and J. L. Durham, "Studies of sulfur compounds absorbed on smoke particles and other solids by photoelectron microscopy," *Proceedings of Air Pollution Control Association Symposium on the Determination of Air Quality, Los Angeles, April, 1971* (Plenum, New York, 1972).
39. S. Beilke and D. Lamb, "Remarks on the rate of formation of bisulfite ions in aqueous solutions," *A.I.Ch.E. J.*, **21**, 402 (1975).
40. J. Müller and S. Beilke, *Abstract of Paper at the Heavy Metal Conference* (Toronto, October, 1975).
41. K. C. Chan, B. L. Cohen, J. O. Frohliger and L. Shabason, "Pittsburgh rainwater analysis by PIXE," *Tellus*, **28**, 24 (1976).
42. H. Ruppert, "Geochemical investigations on atmospheric precipitation in a medium-sized city (Göttingen, F.R.G.)," *Water, Air, Soil Poll.*, **4**, 447 (1975).
43. G. Ketseridis, J. Hahn, R. Jaenicke and C. Junge, "The organic constituents of atmospheric particulate matter," *Atm. Env.*, **10**, 603 (1976).
44. W. W. Blum, E. R. Deane, D. W. Kelin and R. D. Kleopfer, "Sampling and characterization of air for organic compounds," *Water, Air, Soil Poll.*, **4**, 367 (1975).
45. L. A. Barrie, "An experimental investigation of the absorption of sulfur dioxide by cloud and raindrops containing heavy metals," in *Berichte des Institutes für Meteorologie und Geophysik*, No. 28 (University of Frankfurt, Frankfurt a.M., 1975).
46. E. Sawicki *et al.*, *Microchim. Acta*, **5**, 1025 (1968).
47. *Vapor-phase Organic Pollutants* (National Academy of Sciences, Washington, D.C., 1976) pp. 10, 16, 23, 26, 28, 176.
48. B. G. Liptak (editor), *Environmental Engineers' Handbook*, Vol. 2, *Air Pollution* (1974) ch. 3.16, Phenols and Organic Acids.
49. J. R. Smith and J. C. Suggs, "Smoke composition," in *Proceedings of International Symposium on Air Quality and Smoke from Urban and Forest Fires, Colorado State University, Fort Collins, Co.* (National Academy of Sciences, 1976) p. 296, Table 4.
50. S. E. Schwartz and L. Newman, "Processes limiting oxidation of sulfur dioxide in stack plumes," *Env. Sci. Tech.*, **12**, 67 (1978).
51. R. F. Lavrinenko, "The content of sulfur in atmospheric precipitation," *Tr. Glav. Geofiz. Obs.*, **207**, 87 (1968).
52. M. T. Dana *et al.*, *Natural Precipitation Washout of Sulfur Compounds from Plumes*, EPA-R3-73-047 (U.S. Environmental Protection Agency, 1973).
53. Battelle Pacific Northwest Laboratories, *The MAP3S Precipitation Chemistry Network: First Periodic Summary Report* (*September, 1976–June, 1977*), DNL-2402, UC-11, prepared for the U.S. Dept. of Energy under contract EY-76-C-06-1830 (1977).
54. E. R. Altwicker and J. McManus 1977 (unpublished results).
55. U. Hogstrom, "Wet fallout of sulfurous pollutants emitted from a city during rain or snow," *Atm. Env.*, **8**, 1291 (1974).
56. S. Beilke and H. W. Georgii, "Investigation on the incorporation of sulfur dioxide into fog and raindroplets," *Tellus*, **20**, 435 (1968).
57. J. M. Hales, "Fundamentals of the theory of gas scavenging by rain," *Atm. Env.*, **6**, 635 (1972).
58. L. K. Peters, "Some considerations on the washout of sulfate from stack plumes," *Water, Air, Soil Poll.*, **6**, 303 (1976).

59. J. A. Billingsley, S. Kallend and A. R. W. Marsh, *Washout of Sulfur Dioxide by Rain*, Laboratory Note RD/L/N 85/76 (Central Electricity Research Laboratories, Leatherhead, Surrey, U.K., 1976).
60. W. G. N. Slinn, "Some approximations for the wet and dry removal of particles and gases from the atmosphere," *Water, Air, Soil Poll.*, **7**, 513 (1977).

Elmar R. Altwicker
Rensselaer Polytechnic Institute

NITROGEN-DIOXIDE ABSORPTION IN EVAPORATING AND CONDENSING WATER DROPLETS

Introduction

The absorption of gases by clouds, fogs and water droplets is a key step in the removal of many trace gases in the atmosphere. The high acid content of rainfall in many industrial regions is attributed to the absorption of sulfur and nitrogen oxides.[1] In earlier work with single condensing water drops, it was demonstrated that in the case of SO_2, when absorption occurred during droplet growth, the water actually became supersaturated with the trace gas.[2] It is likely that the solution returns to saturation once condensation is stopped but, if dissolved salts are present, they may act to trap the gas in solution and complex it before it has a chance to be released.[3] Similar tests with oxygen as the absorbed gas demonstrated that the mechanism of enhanced mass transfer *via* water-vapor condensation is more general and not exclusively a property of SO_2.[4]

Previous investigations of NO_2 absorption by water have been primarily concerned with fairly concentrated ($>1\%$) amounts, important in the production of nitric acid.[5] At these higher concentrations the nitrogen dioxide is predominantly in the form of the dimer, nitrogen tetroxide

$$2NO_2 \rightleftharpoons N_2O_4, \qquad (1)$$

where $K_{eq} = 3.71$ at 18.8°C.

The overall reaction with water is

$$3NO_2 + H_2O \rightleftharpoons NO + 2HNO_3, \qquad (2)$$

where $K_{eq} = 0.137\ \text{atm}^{-1}$ at 25°C. At higher concentrations ($>10\%$) NO_2 (N_2O_4) may actually react with water vapor in the gas phase before absorption, producing a fine mist.[6] Dekker *et al.* concluded that, at high concentrations, only the N_2O_4 species takes part in the formation of nitric acid.[7]

There are relatively few investigations of water absorption at low concentrations (ppm range) of nitrogen dioxide. Borok determined the Henry's Law constant for NO_2 in water: $H = 100\ \text{atm}^{-1}$.[8] Palmes *et al.*[9] and Crecelius and Forwerg[10] found that the NO_2 is present primarily as NO_2^-, with only trace amounts of NO_3, when absorbed by water in the ppm range. A possible reaction sequence for the low-concentration range may be:

$$3NO_2 + H_2O \rightleftharpoons 2HNO_3 + NO,$$

$$HNO_3 + 2NO + H_2O \rightleftharpoons 3HNO_2,$$

$$HNO_2 \rightleftharpoons H^+ + NO_2^-,$$

where the ionization proceeds to 100% completion as the NO_2 concentration is decreased.

Dekker concluded that the main resistance to mass transfer shifts from the gas phase to the liquid phase as the concentration is lowered. With this background it was decided to test NO_2 in the 100–300-ppm range for enhanced mass transfer *via* water-vapor condensation.

Experimental Arrangement and Procedure

Because of the very low concentrations of NO_2 in water, and since the sample volume is so small, measuring the amount of gas absorbed in water droplets as a function of time demanded an extremely sensitive analytical procedure. It was decided to adopt the Saltzman method, with *N*-(1-naphthyl)ethylene diamine dihydrochloride as the diazotization agent[11]; this method is applicable in the 0.005–5.0-ppm (water) range, and measures the NO_2^- concentration directly.

It was further necessary to design a contact chamber in which the droplets could be exposed to the desired test atmospheres for long enough to ensure that near-saturation conditions could be achieved. It was because of this requirement that we decided to use a suspended drop in a moving gas stream (Figure 1). The contact cell was a glass tube, of length 60 cm and inside diameter 3.5 cm; a 24-gauge hypodermic, with tip polished and flat, was wrapped in Teflon tape, and injected at a point 40 cm from the gas entrance. A droplet 3 mm in diameter was suspended from the capillary at the axis of the tube and was exposed to a predetermined gas mixture flowing at 50.0 cm/s. By manipulating the syringe, a droplet could be released and a new one formed when desired. By synchronizing the droplet injection with a solenoid gas-injection switch, it was possible to assure that the droplet was fully formed and internally at rest before it was exposed to the NO_2 stream. At the commencement of a series of tests, droplets were sampled immediately after injection to ensure that they were free of NO_2^- ions prior to exposure

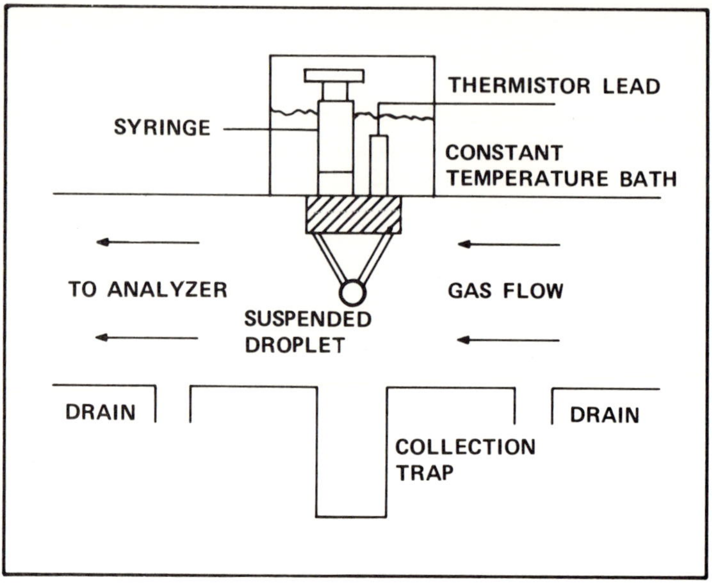

Figure 1 Contact cell.

to the gas stream. The concentrations of NO_2 in the nitrogen tested were 100 and 300 ppm; and the water-vapor pressures were adjusted so that the gas was supersaturated with respect to the initial droplet temperature. Supersaturation ratios (SSR) of 1.0 (no condensation), 1.5, 2.0 and 2.5 were applied at the onset of exposure. Humidity was analyzed with a Cambridge-system Model-880 thermoelectric dew-point hygrometer, and the nitrogen-dioxide content of the gas stream was monitored with a Beckman Model-255A UV analyzer. Water was distilled, deionized and then passed through an oxygen-removal cartridge. The syringe was immersed in a constant-temperature bath at 5°C.

Because of the necessarily long (up to 1 min) exposure times, it was not possible to maintain the droplet at a constant temperature, as was desired. Heating of the droplet occurred both because of the condensation of water vapor from the gas and because of the convective transfer of heat from the warmer air mass (25°C). The droplet temperature was measured with a YSI Model-524 thermistor by suspending the droplet from both the injection capillary and the thermistor needle (see Figure 1). These tests showed that the droplet temperature rises quite rapidly during the first 20 s and then slows to an equilibrium value somewhat lower than the ambient gas temperature. Energy and material balances were written for the droplet and these were solved using the Runge–Kutta approximation.[2] Computer solution shows that the temperature approaches the experimentally determined

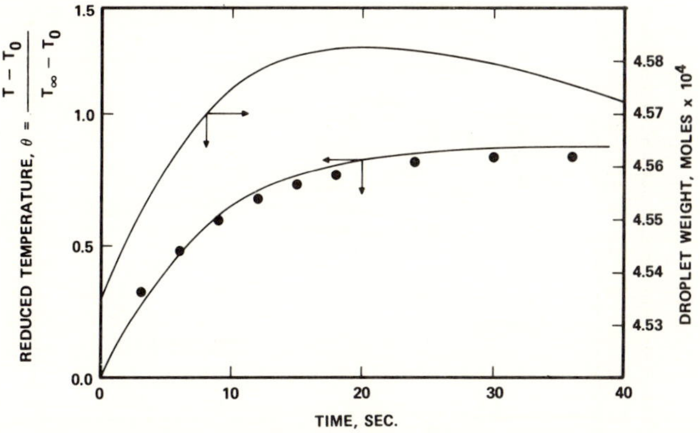

Figure 2 Droplet temperature and mass variation with time. The curves represent Runge–Kutta approximations[2] based on the energy–material balance for a droplet with an initial temperature T_0 = 5°C exposed to flowing air at T_∞ = 25°C and a water-vapor pressure 2.5 times that for water at 5°C. The points are experimentally obtained values.

values quite well. The water vapor condenses until the droplet vapor pressure is the same as that of the ambient gas, and then evaporation begins. Figure 2 shows the case of ambient water-vapor pressure which is 2.5 times that in equilibrium with a 5°C droplet. The total amount of water vapor condensed varied with SSR, and there was only evaporation taking place at an SSR of 1.0.

Results

The results presented in Figures 3 and 4 show the NO_2 absorbed by water as a function of time for various SSR and gas-phase NO_2 concentrations. The results are normalized to the saturation concentration at the temperature of the droplet when sampled. Error bars indicate a ±5% variation in concentration and a ±1-s variation in exposure time. The error in concentration is due mainly to the absorbance reading. A ±1-s error in exposure is attributed to the sampling sequence. For the tests at both 100 and 300 ppm NO_2 in the gas phase, the droplets were practically saturated with NO_2 within 5 s of exposure. The dashed lines indicate the rate of NO_2 (N_2O_4) absorption according to the expression

$$N_{N_2O_4} = (C_i)_{N_2O_4}(Dk)^{1/2}$$

which is based on an assumed first-order reaction during nonsteady-state diffusion into a semiinfinite medium (penetration theory).[12] In this expression, $(C_i)_{N_2O_4}$ is the interfacial concentration of $N_2O_4 = (HP)_{N_2O_4}$; $H_{N_2O_4}$ is the Henry's Law constant; $P_{N_2O_4}$ is the partial pressure of N_2O_4;

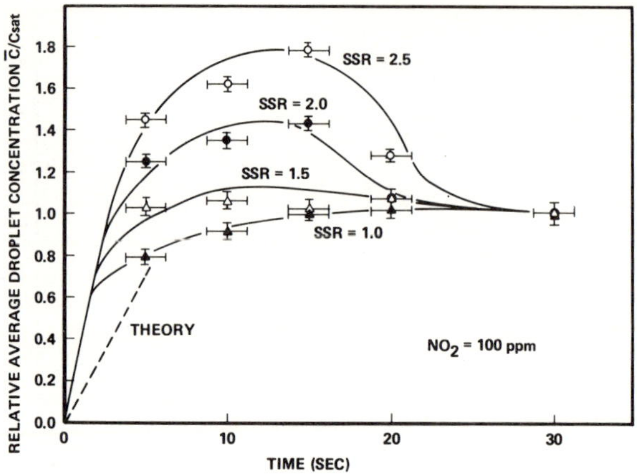

Figure 3 Absorption of 100-ppm nitrogen dioxide from N_2 at various supersaturation ratios (water-vapor condensation rates). The nitrogen–dioxide concentrations are normalized to the saturation value at the droplet temperature. The dashed line represents the theoretical absorption rate based on N_2O_4 data from Kramers *et al.*[13]

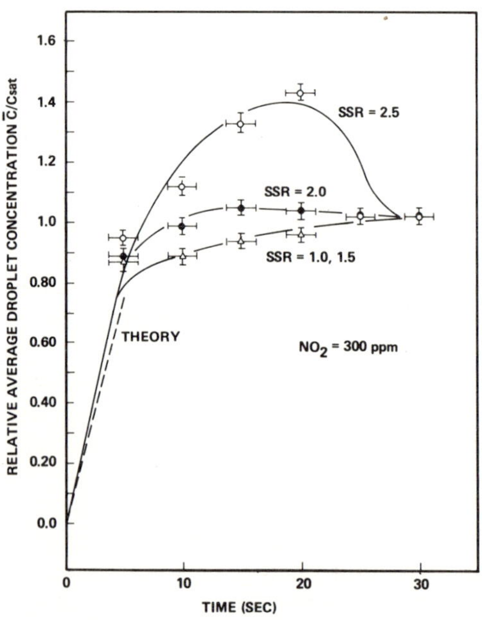

Figure 4 Absorption of 300-ppm nitrogen dioxide from N_2 at various supersaturation ratios (water-vapor condensation rates).

$(H\sqrt{Dk})_{5\,°C} = 8.2 \times 10^{-5}$ g-mole/cm^2·atm·s (based on the results of Kramers *et al.*[13]).

Because the NO_2 concentration in the gas is in the 100–300-ppm range, there should be virtually no N_2O_4 present according to the equilibrium constant expressed earlier. However, as no aqueous absorption data appear to be available in this range, a rough comparison is made between our NO_2 results and those of Kramers *et al.* for N_2O_4,[13] assuming $P_{N_2O_4} = P_{NO_2}$ for convenience. This linear absorption rate in the first few seconds of absorption is somewhat slower than we experienced, which is as would be expected since we are dealing with a condensing system. However, the comparison helps to explain the rather rapid absorption rate in the early (0–5 s) stage of exposure.

There is a definite effect on supersaturation as the SSR is increased above 1.0. This continues for about 15 s and then droplet concentration returns to the saturation level. Tests at 60 s showed that the droplets remain at the saturation concentration. An explanation for the rapid decline in NO_2 supersaturation is indicated in Figure 2, which shows that condensation ceases after 15–20 s and evaporation begins. Therefore, the supersaturation of NO_2 is strongly connected to the condensation of water vapor. For the case of an SSR of 1.0, no condensation occurred but only evaporation. As expected, there is no evidence of supersaturation of NO_2. Greater supersaturation was possible at the lower (100-ppm) than at the higher (300-ppm) NO_2 level, as was experienced in the investigations of SO_2 and O_2.[2,4]

Discussion

The rate of NO_2 absorption during water-vapor condensation or evaporation is a function of the rate of the diffusion-coupled chemical reaction in the liquid phase, mentioned previously, and the rate of water-vapor condensation or evaporation. In order to isolate the effect of the transport of water vapor on the rate of NO_2 absorption, the increase in the amount of nitrogen dioxide absorbed and the corresponding increase in the amount of water condensed for different SSR were determined for fixed values of time. These results indicate that the effect of water-vapor transport on NO_2 absorption follows the relation

$$\frac{d\bar{C}_{NO_2}}{dt} = k(Y_{NO_2})^n \frac{d[H_2O]}{dt},$$

such that

$$\bar{C}_{NO_2} = k(Y_{NO_2})^n[H_2O]$$

where Y_{NO_2} is the NO_2 concentration in the gas phase, expressed as a mole fraction; $[H_2O]$ is the number of moles of water condensed; $k = 3.68 \times 10^{-4}$; $n = 0.105$.

Nitrogen-dioxide absorption is definitely enhanced during water-vapor condensation. If we compare the NO_2 concentration in the condensed-water film with the value calculated from Henry's Law:

$$\frac{\bar{C}_{NO_2}}{[H_2O]} = HY_{NO_2},$$

which, for 100 ppm NO_2 in air at 20°C, becomes

$$\frac{\bar{C}_{NO_2}}{[H_2O]} = 10^{-6},$$

expressed as a mole fraction, whereas in the condensed-water film we get

$$\frac{\bar{C}_{NO_2}}{[H_2O]} = 1.4 \times 10^{-4},$$

again as a mole fraction, which represents supersaturation by a factor of 140. The degree of supersaturation of NO_2 is diminished somewhat as the NO_2 concentration in the gas phase is increased. For instance, a 300 ppm the supersaturation factor is 52.

A possible explanation for this absorption enhancement may be that, under conditions of water condensation, there is a net movement of water vapor toward the water surface, tending to drag extra NO_2 molecules along, whereas in the steady-state condition there exists an equimolar exchange of water molecules both to and from the surface. Nitrogen dioxide is absorbed in both cases, but in the nonsteady state the amount of NO_2 entering the surface is enhanced somewhat by the condensing H_2O molecules. The resistance to absorption is primarily in the liquid phase, which is governed by the NO_2 concentration at the air–water interface. Therefore, if more NO_2 can be packed into the surface, this would tend to enhance the overall absorption process.

The results expressed here show clearly that, as H_2O condensation ceases, the NO_2 concentration level rapidly returns to saturation. There are several mechanisms which may promote this phenomenon: (a) as evaporation takes over, the outer layers of solution are stripped away, releasing NO_2 back into the atmosphere; (b) dissolved NO_2, present either as NO_2^- or NO_3^-, may become too concentrated in the outer layers and revert to NO_2 and then escape as a gas; (c) evaporating water molecules in the vicinity of the interface may reduce the gas-phase NO_2 concentration to a level below that in the bulk gas, providing an extra driving force. That the desorption is so rapid indicates that any one (or a combination) of the above mechanisms is acting more rapidly than the combined liquid-phase-diffusion–chemical-reaction mechanism tending to absorb the NO_2 into the interior of the droplet.

The implication of these results when extended to growing fogs and clouds is that the conversion of ambient NO_2 gas to the dissolved nitrate form may proceed at a much faster pace and to much higher levels than heretofore predicted from accepted absorption theories. Dissolved salts may act to complex the ionized NO_x, thereby trapping it and preventing its release during evaporation.

ACKNOWLEDGEMENTS

This work was supported by Contract EE-77-S-05-5592 from the U.S. Department of Energy.

REFERENCES

1. E. N. Ziegler, "Precipitation composition: Northeastern United States," in *Advances in Environmental Science and Engineering*, Vol. 1, edited by J. R. Pfafflin and E. N. Ziegler (Gordon and Breach, London, 1979).
2. T. L. Wills and M. J. Matteson, "The absorption of sulfur dioxide by water droplets during condensation," in *Colloid and Interface Science*, Vol. 2, edited by Milton Kerker (Academic, New York, 1976) pp. 95–105.
3. M. J. Matteson, W. Stober and H. Luther, **8**, 677, "Kinetics of the oxidation of sulfur dioxide by aerosols of manganese sulfate," *Ind. Eng. Chem.*, **8**, 677 (1969).
4. M. J. Oliver, *The Absorption of Oxygen by Water Droplets During Condensation*, M.S. Thesis (Georgia Institute of Technology, 1976).
5. M. S. Peters, C. P. Ross and J. E. Klein, "Controlling mechanism in the aqueous absorption of nitrogen oxides," *A. I. Ch. E.*, **1**, 105 (1955).
6. H. F. Johnstone and G. C. Williams, "Absorption of gases by liquid droplets," *Ind. Eng. Chem.*, **31**, 993 (1939).
7. W. A. Dekker, E. Snoeck and H. Kramers, "The rate of absorption of NO_2 in Water," *Chem. Eng. Sci.*, **11**, 61 (1959).
8. M. T. Borok, "The dependence of the degree of absorption of nitrogen dioxide in water on its concentration in a gaseous mixture," *Zh. Prikl. Khim.*, **33**, 1761 (1960).
9. E. D. Palmes, A. F. Gunnison, J. Dimattio and C. Tomczyk, "Personal sampler for nitrogen dioxide," *Am. Ind. Hyg. Assoc. J.*, **37**, 570 (1976).
10. H. Crecelius and W. Forwerg, "Investigations of the Saltzman factor," *Staub*, **30**, 23 (1970).
11. B. E. Saltzman, "Colorimetric microdetermination of nitrogen dioxide in the atmosphere," *Anal. Chem.*, **26** (1949).
12. T. K. Sherwood, R. L. Pigford and C. R. Wilke, *Mass Transfer* (McGraw-Hill, New York, 1975).
13. H. Kramers, M. P. Blind and E. Snoeck, "Absorption of nitrogen tetroxide by water jets," *Chem. Eng. Sci.*, **14**, 115 (1961).

John P. Herrmann
and Michael J. Matteson
Georgia Institute of Technology

THE CHEMISTRY AND TRANSPORT OF METHANE AND CARBON MONOXIDE IN THE TROPOSPHERE

1 Introduction

The transport and chemistry of trace species in the troposphere provides one of the most challenging current problems in atmospheric research. Not only are numerous chemical reactions involved, but the complexities of surface–air exchange and troposphere–stratosphere interactions must also be included. A detailed analysis of tropospheric transport and chemistry may never be achieved, but there are some subsets of this system that can be reasonably decoupled from the overall scheme. One such system includes methane, carbon monoxide and the chemical species involved in the conversion of CH_4 to CO in the atmopshere, such as formaldehyde. This article will discuss our present understanding of the physical and chemical behavior of these species in the troposphere. Our presentation will be primarily concerned with those aspects of the CH_4–CO system that relate to the "background" or nonurban atmosphere. It should be noted that, since the discovery of CH_4 and CO in the atmosphere by Migeotte,[1,2] there have appeared several reviews summarizing various aspects of the CH_4–CO system known at the time, including works by Bates and Witherspoon,[3] Ehhalt,[4] Seiler,[5] Wofsy[6] and Jaffe.[7] However, as is so often the case in environmental problems, new atmospheric data and laboratory measurements have necessitated frequent reexaminations of the theories of CH_4 and CO in the atmosphere. In this work we have therefore attempted to present a composite picture of atmospheric CH_4 and CO that reflects the most recent insights and observations.

The CH_4–CO system is somewhat unique in that the processes in which these species are involved are perhaps best considered on a global scale. This implies transformation and exchange processes that occur on a time scale of weeks to months or even years. As such, their importance goes much beyond the urban scale, but at the same time they are not uniformly mixed in the troposphere. It is readily appreciated that the uniformity of trace-species concentrations depends strongly on their atmospheric residence time, and since the residence time of CO is relatively short (on the order of months) its concentration is not very uniform; CH_4 is more uniformly mixed, because its residence time is on the order of years, but measurable spatial and temporal variations apparently still occur.[4]

The global distribution of these gases is not well known at present. It is difficult to obtain an overall view of the distribution by point-sampling methods because of the short residence times; this is especially true for CO. The distribution and concentrations of these species in the atmosphere are coupled by the chemical reactions in which they participate. Figure 1 shows the principal interactions that are significant in this system. In general,

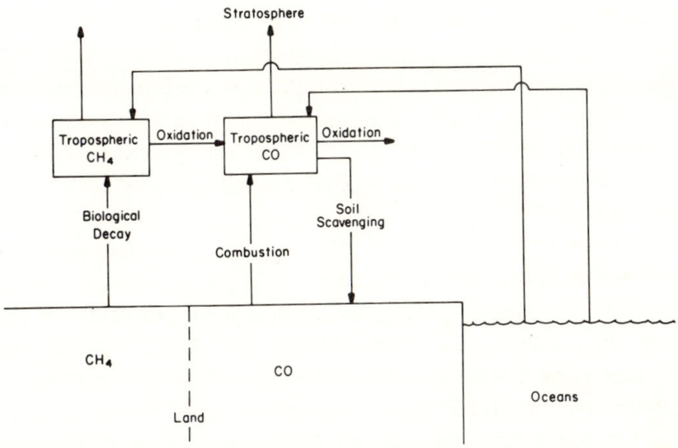

Figure 1 Interactions in the CH_4–CO system. The CH_4 sources are estimated to be 5×10^{14} to 2×10^{15} g/yr; and the sources of CO, excluding that by CH_4 oxidation, are estimated to be in the range $(2–8) \times 10^{14}$ g/yr.

atmospheric oxidation of CH_4 produces H_2CO through a complex sequence of chemical reactions, and the H_2CO further reacts to form CO; this sequence of reactions has key photolytic steps. Other sources and sinks of both compounds contribute to their abundance in the troposphere, and each of these will be discussed in more detail in this article. Furthermore, the magnitudes of these contributions will be considered, and total tropospheric budgets for CH_4 and CO will be presented in Section 9.

While the major thrust of this paper is concerned with nonurban tropospheric CH_4 and CO, such considerations cannot be completely separated from the significant urban contributions; this is particularly true of the impact of anthropogenic CO. Thus, before turning specifically to the nonurban troposphere, some aspects of the urban problem will be briefly summarized.

Emissions of both CH_4 and CO occur within urban regions, although the importance of such CH_4 emissions is not highly significant. As a result, the urban concentration of CH_4 is generally only 50–150% higher than the typical trophospheric concentration of about 1.4 ppm. For example, concentrations in Los Angeles during 1972 averaged only 3.2 ppm.[8] Although CH_4 comprises approximately 40–80% of the total hydrocarbon burden in urban atmospheres, its impact on local air-quality degradation is minor since it does not participate to any significant extent in the reactions important in photochemical-smog episodes.

This minor impact in urban areas is certainly not true for CO. Automotive emissions of CO, resulting from the incomplete combustion of gasoline in internal-combustion engines, account for approximately two-thirds to three-quarters of the anthropogenic emissions,[7,9] which lead to urban CO concentrations much higher than the typical nonurban tropospheric value (about 0.1 ppm). Specifically, the annual mean of CO concentrations in large metropolitan areas can be 30–150 times the typical tropospheric concentration, and the daily maximum concentration may be 100–300 times this base concentration.[8,10]

The CH_4 and CO concentrations resulting from anthropogenic emissions within the urban areas rapidly decrease on moving away from the metropolitan centers. Turbulent mixing and atmospheric transport of the species away from the congestion of these sources quickly dilute the high urban levels to more typical tropospheric levels. The chemical reactions in which these species are involved play a secondary role in this concentration reduction. On the space and time scale of general importance in detailing urban problems, both CH_4 and CO can be classified as chemically inert, and it is common practice in developing air-quality transport models for CO in urban regions to employ the nonreactive assumption. Specifically, the rate of conversion for CO in an urban area is probably less than 0.1%/h or so, justifying the chemically inert classification.

Ironically, while CH_4 and CO photochemical interactions in the highly reactive urban system can often be neglected, it is found that in the nonurban troposphere, where residence times are typically longer, CH_4 and CO play a key role in the photochemistry. As will be discussed below, model calculations indicate that CO and, to a lesser extent, CH_4 abundances help determine the concentration of tropospheric OH, which acts to remove many potentially toxic gases from the atmosphere. As a result, the sources and sinks of CH_4 and CO can indirectly exert a significant influence over our atmospheric environment, and thus attempts to understand and quantify these processes are of much importance.

2 Concentrations of CH_4 and CO in the Troposphere

The presence of CH_4 and CO in the Earth's atmosphere was discovered by Migeotte, who recognized their IR absorption bands in the solar spectrum.[1,2] Subsequent to that discovery, measurement programs have been undertaken to establish the abundance, distribution and variability of CH_4 and CO in the troposphere; to a large extent, these programs are still underway today. The primary measurement technique for CH_4 and CO is gas-chromatographic analysis of grab samples, although Seiler and Junge have developed a continuous CO recording instrument based on HgO–Hg conversion and detection of the Hg atom.[11]

Tropospheric CH_4 appears to be a reasonably well-mixed species with only small spatial and temporal variations in the mixing ratio. Observations indicate a CH_4 abundance of 1.4–1.5 ppm in the Northern Hemisphere, with about 10% lower concentrations in the Southern Hemisphere.[12,13] For example, Wilkniss *et al.*[13] found a CH_4 abundance of 1.44 ± 0.04 ppm between 30°N and 7°N, but only 1.36 ± 0.03 ppm south of 7°N (see Figure 2). This decrease in CH_4 concentration with latitude appears to coincide with the intertropical convergence zone and may be due to enhanced CH_4 production in the Northern Hemisphere as a result of anthropogenic activities such as the cultivation of rice paddy fields and mining operations.[4] Another factor could be the somewhat longer photochemical lifetime of CH_4 in the Northern Hemisphere, believed to be the result of enhanced CO concentrations in the Northern Hemisphere and thus lower OH-radical densities.[6,14] Generally speaking, the relatively constant CH_4 distribution is consistent with a long photochemical lifetime for CH_4—of the order of 10 years.[15] Because the CH_4 abundance exhibits little variability, its average atmospheric abundance

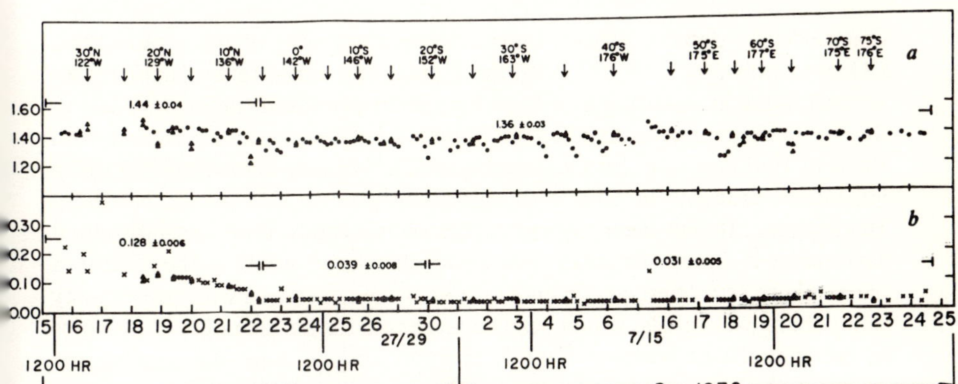

Figure 2 Concentrations of CH_4 (a) and CO (b) over the Pacific Ocean and the Ross Sea, according to Wilkniss *et al.*[13]. (Reprinted with permission from *Nature*.)

has been relatively well defined by the limited observational database accumulated thus far. Therefore, future measurement programs should probably concentrate on establishing more detailed latitudinal and longitudinal trends, as well as identifying any temporal trends that may exist.

In contrast to CH_4, the atmospheric concentration of CO in the atmosphere is highly variable as a function of both position and time. It is therefore very difficult to determine the global CO distribution from the present database, which primarily consists of ground-based observations over land and sea and some airborne data obtained from aircraft and balloons. A review of most of the global CO measurements is available.[5] Figure 3—taken from

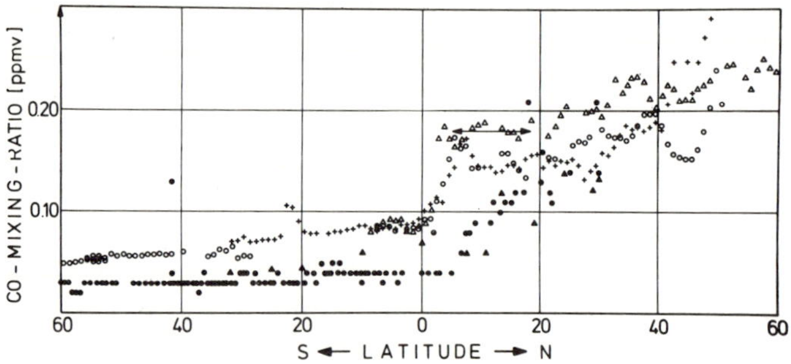

Figure 3 CO mixing ratios in marine air masses over the Atlantic and Pacific Oceans[5]: the filled triangles represent the Eltanin cruise No. 31 (1967) from San Francisco to New Zealand[16]; the open triangles represent the GARP expedition in 1969 from 10°S to 60°N along 30°W over the Atlantic Ocean[11]; the lines represent the average CO mixing ratio found in the Bomex area over the Atlantic Ocean[17]; the open circles represent the Shackleton expedition in November 1971 over the Atlantic Ocean; the crosses represent the Shackleton expedition in April 1972 over the Atlantic Ocean; and the points represent the cruise from the USA to the Antarctic[18] over the Pacific Ocean. (Reprinted with permission from *Tellus*.)

that review[5]—depicts various maritime measurements of the atmospheric CO concentration.[11,16–18,5] With these data, and measurements aboard an aircraft flying at an altitude of 8–10 km, Seiler proposed the average latitudinal CO distribution depicted in Figure 4.[5] These results, coupled with those of Wilkniss *et al.* for CO (see Figure 2),[13] clearly demonstrate a sharp latitudinal gradient in CO, with significantly more CO in the Northern Hemisphere. It has been argued rather convincingly that this latitudinal asymmetry is due to the large anthropogenic source of CO in the Northern Hemisphere.[5,19] These arguments appear to be consistent with present-day photochemical models, which imply a large role for anthropogenic activities in the global CO budget.[6,14,20] It is interesting to note that during the cruise of Wilkniss *et al.*[13] the drop in CO abundance occurred at approximately the same location as the drop in CH_4 levels, possibly implying that the latitudinal gradients of both species are due to the same phenomena.

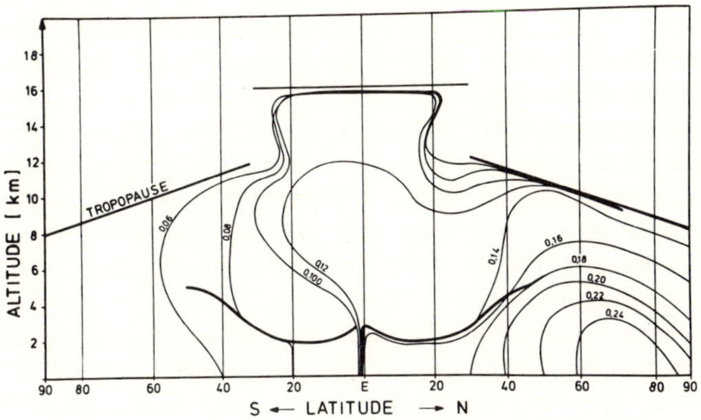

Figure 4 General distribution of CO in marine air in the troposphere and lower stratosphere; the bold lines represent the tropopause and the trade-wind inversion.[5] (Reprinted with permission from *Tellus*.)

Seiler proposes mean tropospheric abundances of 0.15 ppm in the Northern Hemisphere, 0.06 ppm in the Southern Hemisphere and a global average of 0.11 ppm, and concludes that the maximum mean ground-level CO density of about 0.25 ppm occurs between 60°N and 80°N,[5] which is somewhat farther north than might be expected if anthropogenic inputs are the cause of enhanced CO abundances in the Northern Hemisphere. As is mentioned in Section 7, this effect may be due to the large-scale transport of anthropogenic CO. As a result of the high spatial and temporal variability of CO concentrations, the latitudinal distribution proposed by Seiler[5] is still uncertain. Furthermore, very little is known of the longitudinal distribution of CO. Indeed, it is possible that the present Northern Hemispheric CO abundance is heavily weighted by measurements over the Atlantic, which tend to be higher than those over the Pacific Ocean. Thus, when more detailed measurements of CO over the Pacific Ocean and other areas become available, it may be necessary to form a revised picture of the CO distribution. The large spatial and temporal variability of CO makes it an ideal candidate for monitoring from a space platform so as to allow a complete global picture to be accurately assembled.

3 The Oceans as a Source of Methane and Carbon Monoxide

It has long been known that the oceans can act as a repository for atmospheric gases; the most notable example is carbon dioxide. Measurements gathered over the past decade or so imply that the oceans may also act as a source of atmospheric CH_4 and CO. The bulk of this research has been done by two groups—Swinnerton and co-workers at the U.S. Naval Research

Laboratory and Junge, Seiler and co-workers at the Max-Planck Institut für Chemie in West Germany. Both research groups have made simultaneous aqueous-phase and air-phase measurements of the species concentration to investigate this phenomenon. Assuming that both samples are representative of the bulk phases, the relative saturation is established by comparing the data with solubility measurements. In general, the gaseous exchange may be either to the oceans or from the oceans, depending on which phase is supersaturated with respect to the other. Only when the system is at equilibrium does no exchange occur.

In the analytical procedure for measuring the aqueous-phase gas concentration, the first step in the quantitative determination is generally to strip the dissolved gas from the solution. The group at the Naval Research Laboratory then uses selective adsorption in two cold traps to hold the gases; the trapped gases are desorbed from each trap separately and passed to a gas-chromatography unit. Both CH_4 and CO are analyzed using a flame ionization detector, but CO is catalytically reduced to CH_4 before admission to the gas chromatograph. By this procedure, a minimum detection limit of about 10^{-8} ml/liter of solution has been achieved.[21]

Many more measurements of this type have been made for CO than for CH_4. Some of the measurements for both gases are summarized in Figure 5, where ranges of the measured relative saturation are presented. This relative saturation represents the ratio between the aqueous-phase concentration

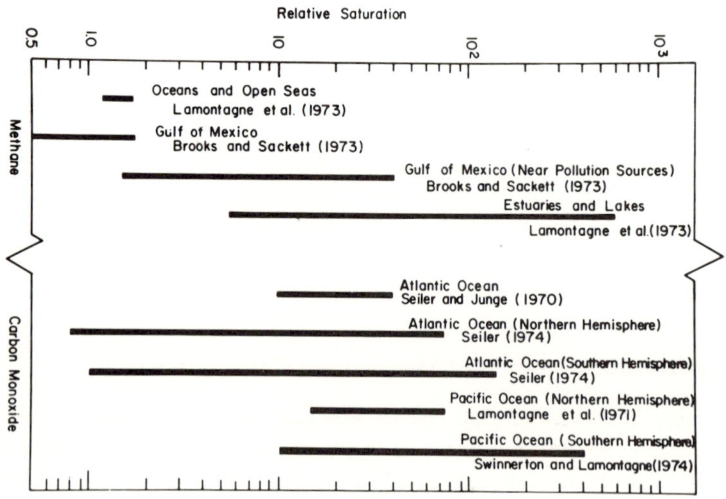

Figure 5 Summary of some of the relative saturation data for CH_4 and CO. A relative saturation greater than 1.0 indicates that the water is supersaturated for that species.

(expressed as a gas-phase concentration equivalent to equilibrium with the aqueous phase) and the concentration in the air phase; a relative saturation greater than unity indicates that the water body is supersaturated in that species. The data shown correspond primarily to surface-layer measurements made during cruises on both the Atlantic and Pacific Oceans.

The supersaturation of CO is considerably higher than that of CH_4, by a factor of 0.8–400, the bulk of the individual measurements being in the range 10–80. An average supersaturation ratio appears to be approximately 35 for CO, which would correspond to an equivalent concentration in the ocean of about 3.5 ppm. The relative saturation of CH_4 is much less, and seems to be approximately 1.2–1.7, corresponding to an equivalent oceanic concentration of around 1.5–2.4 ppm. These values, reported by Lamontagne *et al.*,[22] correspond to surface values for the open ocean. Near-shore, bay and estuary systems can show a considerably higher supersaturation, depending on the time of year, addition of anthropogenic pollutants and amount of tidal mixing. The results specifically cited the following supersaturation ratios for such waters: Chesapeake Bay, 14.3; York River, 21.2; Mississippi River, 5.67; Potomac River, 36.0. These values may also be affected by local pollution. Brooks and Sackett[23] have obtained results for CH_4 in the coastal waters of the Gulf of Mexico which generally support those of Lamontagne *et al.*[22] However, they report that in the Yucutan area, where there is a major upwelling of deep water with low hydrocarbon concentration, the Gulf of Mexico acts as a sink for methane; in this region, they observed the lowest methane relative saturation (about 0.5).

There is strong evidence supporting a diurnal variation of the CO supersaturation in the ocean surface layers, the supersaturation being maximum during the daytime and minimum at night.[18,24] The magnitude of these fluctuations seems to be around ± 40–50% of the mean value. Observations of such behavior on at least two cruises and in both the Atlantic and Pacific Oceans leaves little doubt that the CO supersaturation has a diurnal character; no similar evidence seems to exist for the CH_4 saturation.

Vertical profiles of CH_4 and CO concentrations in natural waters generally show distinctly different patterns. The CH_4 concentration reaches a sharp maximum at a depth of 50–200 m below the surface, and then decreases rapidly.[22,23] This maximum concentration is approximately 2–10 times greater than the surface concentration. The existence of such a maximum would indicate a significant production of CH_4 in this region (possibly of biological origin), the rate of mixing being insufficient to distribute the resulting CH_4 uniformly in the surface layers. Bacterial consumption of methane as it is mixed upwards through aerobic water could also contribute to the vertical gradient in CH_4 concentration. There is probably also a second increase, at depths in excess of 500 m or so.

The concentration of CO in the ocean, on the other hand, decreases continuously with depth.[11,21,25] Below depths of a few hundred meters the concentration seems to be reasonably uniform (10^{-5} ml CO per liter) at 10–50% of the surface value. This type of profile, along with the diurnal variation (suggesting strong dependence on photolytic processes), indicates a near-surface source of marine origin. Laboratory results suggest that dissolved organic carbon may be the source of CO.[26] The same experiments indicate a photolytic dependence for the production of CO, but show no production of saturated hydrocarbons including CH_4.

All of the supersaturations reported for CO were based on solubilities established by Winkler[27] for pure water (0.026 at 288 K) and/or by Douglas[28] for sea water (0.022 at 288 K); both studies used atmospheres of pure CO above the aqueous phases. Meadows and Spedding have suggested that such solubilities may not apply in the atmosphere, since the CO concentration is in the ppm range and lower.[29] As a result, they conducted experiments using distilled water and sea water where the gas-phase CO concentration was in the range 3–18 ppm, and found the solubilities to be more than eight times higher than those reported by Winkler[27] and Douglas.[28] These data would suggest that the oceanic-source strength is not nearly as large as other workers have suggested and that the equilibrium air-phase concentration may be more like 0.3–0.4 ppm. There appears to have been no discussion of the measurements by Meadows and Spedding,[29] but this aspect should certainly be addressed, to clarify the apparent discrepancies.

Determination of the oceanic-source strength of CH_4 and/or CO is a tenuous procedure at best, simply because the data available on the supersaturation and the exchange coefficient between the ocean and air are meager. However, order-of-magnitude estimates can be made, and these will be critically discussed. If the resistance to the transfer from the ocean to the atmopshere may be assumed to be primarily the result of resistance in the ocean phase (i.e., the air-phase resistance is negligible), then the flux J can be written as

$$J = \frac{k_1 c(S - 1)}{H}, \tag{1}$$

where k_1 is the mass-transfer coefficient in the ocean surface layer; c is the concentration of the species in the air; H is the Henry's Law constant describing the phase equilibrium between the atmosphere and ocean; S is the relative saturation. The flux, which can then be multiplied by the ocean surface area to provide the total source strength, depends critically on the values selected for k_1 and S. The values of these parameters used in Eq. (1) should represent some average conditions typical of the phenomena being considered.

First, consider k_l, for which a variety of values have been used. Brtko and Kabel have discussed in detail transfer into water bodies where wind-induced turbulence is the dominant mechanism of mixing in the surface layer.[30] They have correlated the data reasonably well using the wind velocity at a height of 10 m (designated u_{10}) as the independent variable; this is perhaps the best procedure currently available. The obvious requirement is then to select an appropriate value of u_{10}. Three values of k_l based on Brtko and Kabel's analysis are listed in Table I, and these are compared with several other values based on other considerations. These values range from 1.4×10^{-3} to 2.2×10^{-2} cm/s; 5×10^{-3} cm/s is probably a good mean value. Note that, on the basis of this analysis,[30] the value of k_l used by Linnenbom *et al.*[25] would correspond to relatively stiff winds.

TABLE I

Selected values of k_l used for ocean-exchange calculations

Reference	k_l (cm s)
Brtko and Kabel[30]	
$u_{10} = 3$ m/s	1.7×10^{-3}
$u_{10} = 5$ m/s	4.7×10^{-3}
$u_{10} = 10$ m/s	1.8×10^{-2}
Broecker, Li and Peng[31] (used for CO_2 exchange[a])	3.3×10^{-3}
Turbulent film model using the estimate of Okubo[33] for eddy diffusivity (10 cm^2/s) and film thickness of 70 m	1.4×10^{-3}
Linnenbom *et al.*[25]	2.2×10^{-2}
Seiler[5]	9.2×10^{-3}

[a] See also Broecker and Peng.[32]

As already discussed, the supersaturation of CO varies considerably: the bulk of the measurements lie between 10 and 80; 35 probably represents a reasonable mean value.† Nevertheless, this uncertainty in S, along with the uncertainties in k_l, represents a very substantial source of variability in any estimates. These factors must be considered and, accordingly, estimates that represent minimum (based on $k_l = 1.4 \times 10^{-3}$ cm/s and $S = 10$), "best-estimate" ($k_l = 5.0 \times 10^{-3}$ cm/s and $S = 35$) and maximum ($k_l = 1.0 \times 10^{-2}$ cm/s and $S = 80$) values must be provided. Table II gives such estimates for the oceanic strength of CO, along with estimates from several prior studies. The previous estimates are all within the uncertainty of almost two orders of magnitude indicated by the present discussion. It seems that

† Solubilities based on the data of Winkler[27] and Douglas[28] are assumed, since these have been almost exclusively used in other studies.

TABLE II

Estimates of the CO source from the oceans

Probable minimum	5.0×10^{12} g/yr
Best estimate	6.0×10^{13} g/yr
Probable maximum	3.0×10^{14} g/yr
Junge *et al.*[34]	1.6×10^{14} g/yr[a]
Swinnerton *et al.*[35]	9.0×10^{12} g/yr
Linnenbom *et al.*[25]	2.2×10^{14} g/yr[a]
Seiler[5]	1.0×10^{14} g/yr

[a] Adjusted to total ocean surface area.

estimates better than these are simply not possible at the present. Furthermore, improved values for the CO-source strength of the oceans may best be determined by alternative procedures. In any event, as later discussions will show, the oceanic source is small compared to other atmospheric sources.

The source of CH_4 from the oceans is even less certain, but since the supersaturation overall is relatively small it does not apparently represent a significant source. Employing calculations similar to that used for CO would indicate that the upper limit of this source strength is probably around 2×10^{13} g/yr—less than 2–4% of the total biogenic production. Ehhalt has suggested that a typical value is between 4×10^{12} and 6.7×10^{12} g/yr.[4] His typical value for the oceanic source is certainly consistent with the upper limit obtained in the present estimate.

4 Soil–Troposphere Interactions

Present evidence indicates that CH_4 and CO also interact with the Earth's nonocean surface, and the effect of soils and vegetation on the tropospheric concentrations of these species cannot be overlooked. These interactions will be discussed separately for each species.

Substantial quantities of CH_4 are relased into the atmosphere from the Earth's land masses, where it is produced by anaerobic decomposition of organic matter. Ehhalt has discussed these sources in considerable detail and has subdivided them into the following categories: rice paddy fields, swamps and marshes, upland fields, tundra, forests and the enteric fermentation of animals.[4] A summary of these data is shown in Table III, together with the appropriate references. On the basis of these figures, the total biogenic production of CH_4 (excluding the production of oceans, which was discussed earlier) appears to be $(5–8) \times 10^{14}$ g/yr.

TABLE III

Summary of biogenic sources of CH_4

Source	Strength (10^{14} g/yr)	Reference
Rice paddy fields	2.8	Ehhalt[4] update of Koyama[36,37] data
Swamps and marshes	1.2–2.6	Ehhalt,[4] Baker-Blocker *et al.*[38]
Upland fields, forests and tundra	0.12–0.23	Ehhalt[4] estimate based on Koyama[36,37] data
Enteric fermentation of animals	1.0–2.2	Ehhalt[4] update of Hutchinson[39] and Woodwell[40] data

In addition to the biogenic production of CH_4, another $(0.1–0.5) \times 10^{14}$ g/yr are introduced into the atmosphere from volcano activity and anthropogenic sources, such as coal and lignite mining and industrial and automotive emissions.[41] (While direct anthropogenic sources are probably small, the significant emissions due to rice-paddy cultivation and enteric fermentation from cattle are indirectly caused by man's activities.) Ehhalt has suggested that this estimate may be too low by as much as a factor of five, on the basis of the amount of C^{14} in atmospheric CH_4.[4] ($C^{14}H_4$ can be related to biogenic sources.) To justify such a production rate, however, would require a missing nonbiogenic source of CH_4 amounting to about 15–30% of the biogenic source. It is unlikely that such a source has been overlooked, but additional measurements of $C^{14}H_4$ in the atmosphere could prove useful in the further delineation of such disagreements.

While soil and wetland areas are undoubtedly a significant source of CH_4, the present indications are that soils may act as either a source or a sink of CO. Inman and Ingersoll were apparently the first to report on systematic studies to determine the ability of various soils to remove CO from the atmosphere.[42] Between 1971 and 1974, they reported on a series of laboratory and field experiments in which a variety of soils were exposed to test atmospheres containing carbon monoxide.[42–45] Their experiments consisted of exposing soils, either in the laboratory or *in situ* in the field, to test atmospheres initially containing about 100 ppm CO. After taking the proper precautions to prevent leaks from the system, the CO concentration remaining in air above the soil was monitored either by gas chromatography or with an IR gas analyzer. The uptake by the soil was then calculated from the decrease in the CO mixing ratio. Typically, the CO was consumed (at least to the detection limit of their analytical procedure) within a few hours, and an average uptake rate was then obtained.

The principal findings of their experiments can be summarized as follows.

a) The uptake rates in laboratory tests at 298 K varied from 6.1×10^{-11} to 4.7×10^{-10} g/cm^2/s. In field experiments the uptake rates, corrected for seasonal temperature fluctuations, ranged from 2.1×10^{-10} to 3.2×10^{-9} g/cm^2/s.

b) There was a strong effect of temperature on the uptake rate, which was a maximum around 303 K and decreased to approximately 10% of the peak value at temperatures below 283 K or greater than 323 K.

c) Soil moisture content had relatively little effect on the uptake rate, but soils which had a high organic content generally resulted in increased uptakes. Furthermore, soils under cultivation typically had lower uptake rates than similar soils under natural vegetation; this was attributed to reduced organic matter in the surface layers of the cultivated soils.

d) CO assimilation in the soils appeared to be of biological origin; fungi were believed to be the principal active type of species. Sterilization by autoclaving completely eliminated the activity, whereas the addition of small amounts of nonsterile soil to previously autoclaved soil (1 g added to 2.8 kg) ultimately led to the expected CO uptake rates.

e) Apparently, soils could be conditioned to provide increased uptake rates. Specifically, roadside soils were consistently higher in CO-uptake capacity than similar soils from remote areas. Laboratory experiments indicated that this increased rate was due to the constant exposure of roadside soils to high CO concentrations.

Perhaps the most significant criticism of the work of Ingersoll, Inman and co-workers is the high CO concentrations in the test atmospheres above the soils. The typical concentration of 100 ppm used is much higher than the expected nonurban CO concentration in the troposphere. As a result, it is necessary to find procedures by which their results may be adjusted to more typical tropospheric concentrations. It seems reasonable to assume that the action of the soils on atmospheric CO is a first-order process with respect to CO, and that the rate of uptake depends on the area of soil exposed. On this basis then, the exchange would be analogous to that between the oceans and atmosphere, although the values of the parameters describing the process would probably be different. If such a model is employed to analyze their results, an exponential decrease in the CO concentration would be anticipated in their experiments; i.e.,

$$\frac{c}{c_0} = \exp\left(-k_s \frac{A}{V} t\right), \qquad (2)$$

where c is the concentration of CO in the test atmosphere at time t; c_0 is the initial concentration of CO in the test atmosphere; k_s is the soil–air exchange coefficient, which depends on the type of soil; A is the superficial area of the exposed soil; V is the volume of the test atmosphere; t is the time from the start of the experiment. Equation (2) contains two implicit assumptions: (a) that there is no resistance to the exchange process in the gas phase; and (b) that there is sufficient soil for all of the CO in the test atmosphere to be assimilated. The parameter k_s could be obtained from the slope of a plot of ln (c/c_0) versus t if adequate data were available. However, an alternative procedure is to use an appropriate average CO concentration in the test atmosphere. On the basis of the model, a logarithmic average concentration would be the most suitable, and therefore an average CO concentration of 20–30 ppm would be reasonable. Taking a value of 25 ppm, the uptake exchange coefficient, k_s, for the Ingersoll–Inman data would be from 2.1×10^{-3} to 1.6×10^{-2} cm/s for the laboratory experiments, and from 7.1×10^{-3} to 1.1×10^{-1} cm/s for the field experiments. These values thus normalize the extensive data of Ingersoll, Inman and co-workers, and can then be used as estimates of the capacity of the soils as a sink for CO.

Liebl also conducted a series of experiments similar to those of Ingersoll, Inman and co-workers, except that the initial CO mixing ratio was 0.2–0.3 ppm—a much more realistic value.[46] His measured uptake rates also showed a temperature dependence, although a decrease in uptake rate at temperatures below 293 K was not observed. In fact the maximum uptake rate shown was at 283 K, the lowest temperature for which data were presented. At 288 K, an uptake rate averaged over all types of tested soils was 1.5×10^{-11} g/cm^2/s. This would correspond to $k_s \simeq 5 \times 10^{-2}$ cm/s, a number close to the midpoint of the values calculated for the Ingersol–Inman field experiments. Thus, this procedure shows that the data from the two groups of investigators are consistent, at least to rough estimates.

Liebl's experiments did show one interesting aspect that the experiments of Ingersoll, Inman and co-workers at the higher CO concentrations did not reveal: there is an apparent equilibrium between CO-production and CO-destruction mechanisms.[46] When using test air initially free of CO, an increase in the CO concentration was observed. The final equilibrium value was essentially the same as that observed for the same temperature when starting with CO-rich air; the single set of data reported in the review of Liebl's work by Seiler[5] indicates that this equilibrium value is around 0.17 ppm. Although the Ingersoll–Inman studies did not report an analogous equilibrium value, they did report evolution of CO from soil samples that had previously been autoclaved for sterilization experiments.[43]

It would obviously be difficult to estimate the soils' sink potential on a global scale on the basis of these results. Such an estimate is confounded by the equilibrium value suggested by Liebl's results. In fact, if the equilibrium soil activity corresponds to a mixing ratio greater than about 0.15 then the soils can act as a net source of CO. The converse would, of course, be indicated for mixing ratios less than that value. This is a point that should be closely considered and needs clarification through additional experiments. Nevertheless, some estimates of the sink potential can be made by assuming the equilibrium value to be zero and using appropriate values of k_s. On this basis, and using an atmospheric CO mixing ratio of 0.1 ppm, this sink potential would probably be in the range from 3×10^{13} to 5×10^{14} g/yr, with 2×10^{14} g/yr as the "best-estimate" value; Seiler previously suggested a value close to the upper end of this range.[5] It must be emphasized again, however, that if there is an equilibrium soil activity then the soils could represent a net CO source of roughly comparable magnitude. On the other hand, the results of Ingersoll, Inman and co-workers suggest that soils—at least certain types—are able to adapt to higher CO mixing ratios by increasing their uptake rate. Thus, the soils may possibly afford a substantial buffering capacity against increased tropospheric CO concentrations.

While the anthropogenic production of CH_4 appears to be small, emissions of CO at ground level due to anthropogenic activities are quite significant. CO is produced in the incomplete combustion of organic material, and it is believed that more than 70% of the anthropogenic emissions are related to mobile sources.[47] Seiler states that 85% of the anthropogenic production of CO occurs in the Northern Hemisphere.[5] Calculations of global anthropogenic CO production rates involve correlating data on world usage of fossil fuels with CO emission factors for various combustion processes, and the results are highly uncertain. Estimates by other investigators are listed in Table IV. The upward trend in these rates reflects both the increasing production of CO over the past 25 years and the more detailed and complete analyses of CO-source strengths carried out by the later studies. Seiler's estimate for 1973 of 6.4×10^{14} g/yr includes globally averaged CO emission factors, which tend to be somewhat larger than emission factors for the United States. The probable present and future trends in global CO emissions and the effects of pollution control in the United States are discussed in

TABLE IV

Estimates of anthropogenic production rates

Source	Year	Rate (10^{14} g/yr)
Bates and Witherspoon[3]	1952	2.1
Robinson and Robbins[16]	1966	2.9
Jaffe[7]	1970	4.1
Seiler[5]	1973	6.4

Section 8. Note, however, that an increase in global emissions at a rate of 1.5%/yr from 1973 onward could lead to a CO production of as much as 6.9×10^{14} g/yr by the end of 1978. Presently, anthropogenic CO emissions are most probably about (5–7) $\times 10^{14}$ g/yr, but may possibly differ from this range by as much as a factor of two.

5 Transport of CH_4 and CO to the Stratosphere

Stratospheric air is generally characterized by lower concentrations of CH_4 and CO than air in the troposphere. Thus, the continuous exchange of air between troposphere and stratosphere implies a net flux of CH_4 and CO from the troposphere to the stratosphere, the stratosphere being a net sink for both species. The magnitude of the tropospheric–stratospheric exchange of CH_4 and CO has been estimated using three approaches: (a) dynamical studies of air transport between the troposphere and stratosphere; (b) considering the combined CH_4–CO chemistry; and (c) analysis of species concentration profiles to deduce fluxes at the tropopause. The results obtained with each approach will be briefly reviewed, and then the potential impact of perturbations of the CH_4 and CO concentrations in the troposphere on stratospheric processes will be discussed.

Newell *et al.* used the first approach to deduce the CO flux associated with the exchange of air across the tropopause.[48] At tropical latitudes, rising large-scale motions result in a flux of air into the stratosphere, while at middle latitudes downward motions cause a corresponding return air flux into the troposphere. The net CO transport accompanying these movements depends on the difference between the CO concentration at the tropical tropopause and that in the lower stratosphere of the middle latitudes. Thus, suitable CO concentrations in these regions are required to complete the analysis. Newell *et al.* superposed the effect of CO transport by eddy processes on this bulk exchange, and their estimate based on both processes was approximately 1×10^{13} g CO/yr.[48] This estimate is given in Table V, together with the other estimates for CH_4 and CO discussed below.

Pressman and Warneck used similar considerations in combination with time constants to investigate interhemispheric exchange and troposphere–stratosphere exchange.[49] They concluded that the stratosphere could consume 11% per year of the total CO inventory of the stratosphere, which corresponds to about 6×10^{13} g/yr.

Ehhalt and Heidt used a procedure analogous to that of Newell *et al.*[48] for CH_4, by considering the Hadley-cell circulation which penetrates into the lower stratosphere, and then including upward eddy-diffusion transport.[4,12] These considerations, in combination with the assumption that all the CH_4 entering the stratosphere is destroyed, led to the conclusion that this effect

TABLE V

Data on transport from the troposphere to stratosphere

Investigator	Approach	Transport CH_4	(10^{14} g/yr) CO
Pressman and Warneck[49]	Transport, residence time	—	0.6
Wofsy *et al.*[50] (also McConnell *et al.*[51])	Chemistry	0.17	0.3
Machta[52]	Profile	0.5	1.9
Ehhalt and Heidt[12] (also Ehhalt[4])	Transport	1.0–2.0	—
Newell *et al.*[48]	Transport	—	0.1
Seiler[5]	Profile	—	1.1

corresponds to a sink of (1–2) × 10^{14} g/yr.[12] However, since apparently no account was taken of the return flux of air into the troposphere from the stratosphere, which would carry CH_4 at mixing ratios of about 1 ppm, this value must be regarded as an upper limit and may be as much as a factor of three too high.

Turning to the second approach, a number of papers have used the detailed chemistry of CH_4 and CO to estimate the net fluxes of these two species in the atmosphere at any altitude.[50,51,53] Typical of these calculations is that of Wofsy *et al.*[50] who employed a one-dimensional vertical model which coupled vertical transport (parameterized using an eddy diffusion coefficient) and the detailed temperature-dependent photochemistry. In a steady state, the flux of CH_4 and CO to the stratosphere is equal to the net photochemical-column loss rate of CH_4 and CO above the tropopause. Assuming a vanishing flux at 50 km for CH_4 and at 120 km for CO, this approach yielded fluxes into the stratosphere of 1.7 × 10^{13} g CH_4/yr and 3 × 10^{13} g CO/yr. Their estimate for CO leakage to the stratosphere is in remarkably good agreement with the estimate obtained by Newell *et al.* from transport considerations.[48]

The profile method—the third approach—utilizes the concentration profile of a species in combination with estimates of the eddy diffusivity to establish the flux J; i.e.,

$$J = -K_z \frac{\partial x}{\partial z}\bigg|_{\text{tropopause}}, \tag{3}$$

where K_z is the vertical eddy diffusion coefficient; x is the appropriate mixing ratio; z is the altitude. Machta employed profiles from ground level to the lower stratosphere, and thus required values of K_z for both the troposphere and stratosphere.[52] He took values of 6 × 10^4 cm²/s and 5 × 10^3 cm²/s, respectively, and obtained estimates for both CH_4 and CO. On the other hand, Seiler, using an analogous technique for CO, took its concentration to be constant at the tropopause and set $K_z = 3 \times 10^3$ cm^2/s for the stratosphere.[5] Machta reported annual losses to the stratosphere of 5×10^{13} g CH_4 and 1.9×10^{14} g CO,[52] whereas Seiler's estimate for CO was 1.1×10^{14} g.[5] Note that these two estimates for CO transport would be nearly identical if both investigators had used the same value for the vertical eddy diffusivity in the stratosphere. Furthermore, these are the highest estimates for the CO losses to the stratosphere that are listed in Table V. This is most probably because losses by chemical reactions at any level are not accounted for in this procedure, and hence the total gradient of the species concentration is presumed to be due only to eddy-diffusion transport; these estimates[52,5] must therefore be regarded as upper bounds. Furthermore, the difficulties associated with establishing representative values for the concentration gradient and K_z at the tropopause make this method somewhat uncertain; in fact, the profile method is very approximate, as can be seen by comparing these values with those of Wofsy *et al.*[50]

Overall consideration of all the results seems to indicate the following ranges for the transport of CH_4 and CO from the troposphere into the stratosphere: $(1–7) \times 10^{13}$ g GH_4/yr and $(1–6) \times 10^{13}$ g CO/yr.

The oxidation of the CH_4 and CO that is transported into the stratosphere is believed to have an effect on the stratospheric photochemistry. In addition to oxidation by OH

$$CH_4 + OH \longrightarrow CH_3 + H_2O,$$

stratospheric CH_4 reacts with $O(^1D)$

$$CH_4 + O(^1D) \longrightarrow CH_3 + H_2O,$$

$$CH_4 + O(^1D) \longrightarrow H_2 + H_2CO.$$

The increased intensity of UV radiation in the stratosphere results in a higher concentration of OH radicals and $O(^1D)$ atoms, which, when coupled with the positive temperature gradient over the altitude, leads to a decrease in the photochemical lifetime of CH_4 in the stratosphere. Calculations yield lifetimes of about 10^2 yr at 10 km, 3 yr at 30 km, and 0.2 yr at 40 km. This decrease in photochemical lifetime, coupled with the slow rate of transport through the tropopause, results in a decreasing CH_4 mixing ratio with altitude in the stratosphere and mesosphere. Figure 6 (from Liu *et al.*[54]) illustrates a typical CH_4 profile obtained from one-dimensional model calculations; the H_2O and H_2 profiles are included for comparison. As in the troposphere, the production of CH_3 from the oxidation of CH_4 leads to a reaction sequence known as the methane-oxidation chain—discussed in detail in Section 6—and ultimately produces CO, which is oxidized to CO_2.

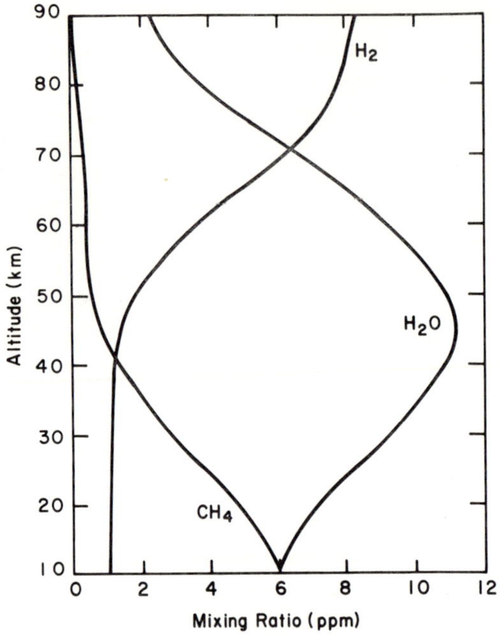

Figure 6 Calculated mixing ratios of (2 × H_2O), (2 × H_2), and (4 × CH_4) as a function of the altitude.[59]

One of the major effects of CH_4 degradation in the stratosphere is the production of H_2O. Since the transport of H_2O from the troposphere to the stratosphere is partially blocked by the colder temperatures at the tropopause, the photochemical production of H_2O can be quite significant. The importance of stratospheric H_2O production is illustrated in Figure 6, which indicates that the decrease in CH_4 with altitude is complemented by a corresponding increase in H_2O in the stratosphere. In this graph, approximately 50% of the H_2O in the upper stratosphere is from CH_4 degradation.

As a source of stratospheric H_2O, CH_4 indirectly leads to the production of free radicals; i.e.,

$$H_2O + O(^1D) \longrightarrow 2OH,$$

and thus affects the stratospheric O_3 abundance through reactions such as

$$OH + O_3 \longrightarrow HO_2 + O_2,$$

and

$$HO_2 + O_3 \longrightarrow OH + 2O_2.$$

On the other hand, CH_4 can act to enhance O_3 by reacting with Cl

$$CH_4 + Cl \longrightarrow CH_3 + HCl,$$

and thus terminating the Cl–ClO catalytic chain

$$Cl + O_3 \longrightarrow ClO + O_2,$$

$$O_3 + h\nu \longrightarrow O_2 + O,$$

$$ClO + O \longrightarrow Cl + O_2.$$

The Cl–ClO catalytic chain destroys stratospheric O_3.[55] Recent model calculations imply that the role of CH_4 as a Cl–ClO chain terminator dominates over its role as a source of free radicals,[56,57] and thus an increase in stratospheric CH_4 will tend to increase stratospheric O_3 concentrations.

The oxidation of stratospheric CO can also affect stratospheric O_3 through the following reaction sequence

$$CO + OH \longrightarrow CO_2 + H,$$

$$H + O_2 + M \longrightarrow HO_2 + M,$$

$$NO + HO_2 \longrightarrow NO_2 + OH,$$

$$NO_2 + h\nu \longrightarrow NO + O,$$

$$O + O_2 + M \longrightarrow O_3 + M,$$

$$\text{Net:}\quad CO + 2O_2 + h\nu \longrightarrow CO_2 + O_3.$$

Since the tropospheric CH_4 and CO abundances determine the levels of CH_4 and CO in the stratosphere, a perturbation in tropospheric CH_4 and CO may lead to changes in stratospheric O_3. A complete discussion of the implications of a change in the atmospheric CH_4 and CO abundances is given in Section 8.

6 Photochemical Production and Destruction

The role of photochemistry in the budgets of CH_4 and CO was first outlined in works by Weinstock,[58] Levy [59] and McConnell *et al.*[51] On the basis of these studies and more recent work, Table VI lists the reactions and the rate constants generally included in studies of tropospheric photochemistry and its relation to atmospheric CH_4 and CO.[50,53,59–64] Although known values of the rate constants are used where possible, not all of the assigned rate constants and branching ratios have been well determined. The values listed in Table VI represent judgements based on the present knowledge, and some of these will undoubtedly be modified in the future.

TABLE VI

Rate coefficients of important tropospheric chemical reactions

	Reaction	Rate coefficient[a]	Reference
R1a	$CH_4 + OH \rightarrow CH_3 + H_2O$	$K_{1a} = (2.35 \times 10^{-12})\exp(-1710/T)$	Hudson[65]
R1b	$CH_4 + O(^1D) \rightarrow CH_3 + OH$	$K_{1b} = 1.3 \times 10^{-10}$	Hudson[65]
R2	$CH_3 + M + O_2 \rightarrow CH_3O_2 + M$	$K_2 = 3 \times 10^{-32}$	Heicklen[66]
R3a	$CH_3O_2 + HO_2 \rightarrow CH_3OOH + O_2$	$K_{3a} = K_{19}$	Levy[67]
R3b	$2CH_3O_2 \rightarrow 2CH_3O + O_2$	$K_{3b} = 2.6 \times 10^{-13}$	Heicklen[66]
R4a	$CH_3OOH + OH \rightarrow CH_3O_2 + H_2O$	$K_{4a} = K_{21}$	Levy[67]
R4b	$CH_3OOH \rightarrow$ Rainout	$K_{4b} = 1 \times 10^{-6}$	Assumed
R5	$CH_3OOH + h\nu \rightarrow CH_3O + OH$	$K_5 = K_{20}$	Levy[67]
R6a	$NO + CH_3O_2 \rightarrow H_2CO + HNO_2$	$K_{6a} = 0$	Pate *et al.*[68]
R6b	$NO_2 + CH_3O_2 \rightarrow H_2CO + HNO_3$	$K_{6b} = 0$	Assumed
R6c	$NO + CH_3O_2 \rightarrow CH_3O + NO_2$	$K_{6c} = (3.3 \times 10^{-12})\exp(-500/T)$	Demerjian *et al.*[69]
R7	$CH_3O + O_2 \rightarrow H_2CO + HO_2$	$K_7 = (1.6 \times 10^{-13})\exp(-3300/T)$	Hudson[65]
R8	$H_2CO + OH \rightarrow HCO + H_2O$	$K_8 = (3 \times 10^{-11})\exp(-250/T)$	Hudson[65]
R9	$H_2CO + h\nu \rightarrow HCO + H$	$J_9(Z = 0) = 7.1 \times 10^{-6}$ [b]	Calvert *et al.*[70]
R10	$H_2CO + h\nu \rightarrow CO + H_2$	$J_{10}(Z = 0) = 2.2 \times 10^{-5}$ [b]	Calvert *et al.*[70]
R11	$H + O_2 + M \rightarrow HO_2 + M$	$K_{11} = (3 \times 10^{-32})(273/T)^{1.3}$	Hudson[65]
R12	$HCO + O_2 \rightarrow CO + HO_2$	$K_{12} = 1 \times 10^{-13}$	McMillan and Calvert[71]
R13a	$CO + OH \rightarrow CO_2 + H$	$K_{13a} = (2.1 \times 10^{-13})\exp(-115/T) + (7.3 \times 10^{-33})n_M$	Chan *et al.*[72]
R13b	$CO + HO_2 \rightarrow CO_2 + OH$	$K_{13b} = 1 \times 10^{-20}$	Davis *et al.*[73]
R14	$H_2 + OH \rightarrow H + H_2O$	$K_{14} = (6.8 \times 10^{-12})\exp(-2020/T)$	Greiner[74]
R15	$O(^1D) + H_2O \rightarrow 2OH$	$K_{15} = 2.3 \times 10^{-10}$	Davidson *et al.*[75]
R16	$O(^1D) + H_2 \rightarrow H + OH$	$K_{16} = 1.3 \times 10^{-10}$	Davidson *et al.*[75]
R17	$HO_2 + OH \rightarrow H_2O + O_2$	$K_{17} = 3 \times 10^{-11}$	Hudson[65]
R18	$OH + OH \rightarrow H_2O + O$	$K_{18} = (1 \times 10^{-11})\exp(-550/T)$	Baulch *et al.*[76]
R19	$HO_2 + HO_2 \rightarrow H_2O_2 + O_2$	$K_{19} = (3 \times 10^{-11})\exp(-500/T)$	Hampson *et al.*[77]
R20a	$H_2O_2 + h\nu \rightarrow 2OH$	$J_{20a}(Z = 0) = 2.2 \times 10^{-6}$ [b]	Molina *et al.*[78]
R20b	$H_2O_2 \rightarrow$ Rainout	$K_{20b} = 1 \times 10^{-6}$	Assumed
R21	$H_2O_2 + OH \rightarrow HO_2 + H_2O$	$K_{21} = (1 \times 10^{-11})\exp(-750/T)$	Hudson[65]
R22	$HO_2 + NO \rightarrow OH + NO_2$	$K_{22} = 8 \times 10^{-12}$	Howard and Evenson[79]

R22a	$NO_2 + HO_2 + M \rightarrow HOONO_2 + M$	$K_{22a} = 1.97 \times 10^{-31}$	Howard[80]
R22b	$HOONO_2 \rightarrow HO_2 + NO_2$	$K_{22b} = (1.4 \times 10^{14})\exp(-10350/T)$	Graham *et al.*[81]
R23	$OH + NO_2 \xrightarrow{M} HNO_3$	$K_{23} \cong 1.25 \times 10^{-11}$	Hudson[65]
R24	$OH + NO \xrightarrow{M} HNO_2$	$K_{24} = 2 \times 10^{-12}$	Baulch *et al.*[76]
R25a	$HNO_3 + h\nu \rightarrow OH + NO_2$	$J_{25a}(Z = 0) = 1.1 \times 10^{-7}$ [b]	Johnston and Graham[82]; a quantum yield of 1.0 was assumed
R25b	$HNO_3 + OH \rightarrow NO_3 + H_2O$	$K_{25b} = 8 \times 10^{-14}$	Hudson[65]
R26	$HNO_3 \rightarrow$ Rainout	$K_{26} = 1 \times 10^{-6}$	Assumed
R27	$HNO_2 + h\nu \rightarrow OH + NO$	$J_{27}(Z = 0) = 1.5 \times 10^{-4}$ [b]	Johnston and Graham[82]
R28	$NO_2 + O_3 \rightarrow NO_3 + O_2$	$K_{28} = (1.1 \times 10^{-13})\exp(-2450/T)$	Garvin and Hampson[83]
R29	$NO + O_3 \rightarrow NO_2 + O_2$	$K_{29} = (2.1 \times 10^{-12})\exp(-1450/T)$	Hudson[65]
R30	$NO_2 + h\nu \rightarrow NO + O$	$J_{30}(Z = 0) = 3.8 \times 10^{-3}$ [b]	Hampson *et al.*[77]
R31a	$NO_3 + NO_2 \xrightarrow{M} N_2O_5$	$K_{31a} = 3.8 \times 10^{-12}$	Baulch *et al.*[76]
R31b	$NO_3 + NO_2 \rightarrow NO + NO_2 + O_2$	$K_{31b} = (2.3 \times 10^{-13})\exp(-1000/T)$	Garvin and Hampson[83]
R32	$NO_3 + NO \rightarrow 2NO_2$	$K_{32} = 8.7 \times 10^{-12}$	Harker and Johnston[84]
R33a	$NO_3 + h\nu \rightarrow NO_2 + O$	$J_{33a}(Z = 0) = 2.3 \times 10^{-3}$ [b]	Cross sections taken from Johnston and Graham[82]; a 10% quantum yield for each process was assumed
R33b	$NO_3 + h\nu \rightarrow NO + O_2$	$J_{33b}(Z = 0) = 2.3 \times 10^{-3}$ [b]	
R34	$NO + NO_2 + H_2O \rightarrow 2HNO_2$	$K_{34} = 6 \times 10^{-37}$	Chan *et al.*[85]
R35	$N_2O_5 + H_2O \rightarrow 2HNO_3$	$K_{35} = 1 \times 10^{-20}$	Morris and Niki[86]
R36a	$N_2O_5 \rightarrow NO_3 + NO_2$	$K_{36a} = (5.7 \times 10^{14})\exp(-10600/T)$	Baulch *et al.*[76]
R36b	$N_2O_5 + h\nu \rightarrow NO_2 + NO_3$	$K_{36b} = 2.0 \times 10^{-5}$	Levy[67]
R37	$O_3 + h\nu_{visible} \rightarrow O_2 + O$	$J_{37}(Z = 0) = 1.3 \times 10^{-4}$ [b]	Griggs[87]; Kuis *et al.*[88]
R38	$O + O_2 + M \rightarrow O_3 + M$	$K_{38} = (1.1 \times 10^{-34})\exp(510/T)$	Huie *et al.*[89]
R39	$O_3 + h\nu_{UV} \rightarrow O(^1D) + O_2$	$J_{39}(Z = 0) = 3.8 \times 10^{-6}$	Griggs[87]; Kuis *et al.*[88]
R40	$O(^1D) + M \rightarrow O + M$	$K_{40} = 3.2 \times 10^{-11}$	Davidson *et al.*[75]
R41	$OH + O_3 \rightarrow HO_2 + O_2$	$K_{41} = (1.5 \times 10^{-12})\exp(-1000/T)$	Hudson[65]
R42	$HO_2 + O_3 \rightarrow OH + 2O_2$	$K_{42} = (1.0 \times 10^{-13})\exp(-1250/T)$	Garvin and Hampson[83]
R43	$CH_3Cl + OH \rightarrow CH_2Cl + H_2O$	$K_{43} = (2.2 \times 10^{-12})\exp(-1142/T)$	Hudson[65]
R44	$CH_3CCl_3 + OH \rightarrow CH_2CCl_3 + H_2O$	$K_{44} = (3.5 \times 10^{-12})\exp(1562/T)$	Hudson[65]

[a] Units: s^{-1} for one-body reactions, cm^3/s for two-body reactions and cm^6/s for three-body reactions.
[b] The quantities shown are the daily averaged photodissociation frequencies for equinoctial conditions at 45°N.

The CH_4 and CO photochemistry is initiated by the production of free radicals *via* the photolysis of O_3 by near-UV photons

$$O_3 + h\nu \longrightarrow O_2 + O(^1D), \qquad (R39)$$

followed by

$$O(^1D) + H_2O \longrightarrow 2OH. \qquad (R15)$$

The OH radicals may then oxidize CH_4

$$CH_4 + OH \longrightarrow CH_3 + H_2O, \qquad (R1)$$

leading to the methane-oxidation chain. This reaction chain, illustrated in Figure 7, eventually results in the production of H_2CO, H_2 and CO. The oxidation of CO by OH

$$CO + OH \longrightarrow CO_2 + H \qquad (R13a)$$

terminates the reaction chain; at this point, a CH_4 molecule has been converted to CO_2. Reactions (R1) and (R13a) are believed to represent major loss processes for atmospheric CH_4 and CO, respectively, while the production of CO from the CH_4 oxidation may be an important natural source of atmospheric CO. In this section the magnitude of these sources and sinks is estimated.

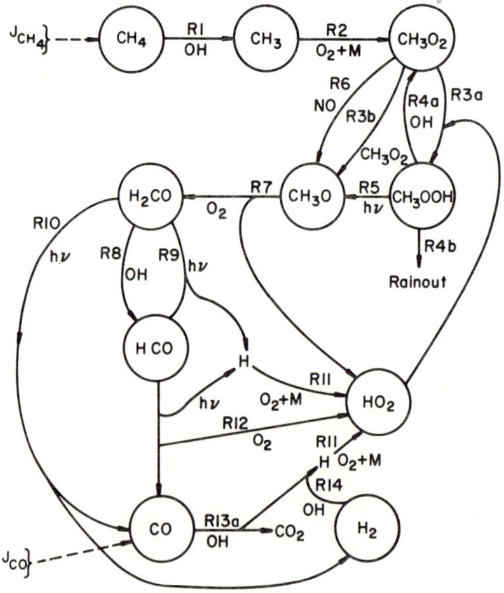

Figure 7 The CH_4 oxidation chain; J_{CH_4} and J_{CO} represent the ground sources of CH_4 and CO, respectively.

The OH radicals consumed by Eqs. (R1) and (R13a), as well as by

$$H_2CO + OH \longrightarrow HCO + H_2O \qquad (R8)$$

and

$$H_2 + OH \longrightarrow H + H_2O, \qquad (R14)$$

lead to the production of HO_2 radicals *via* the reactions

$$HCO + O_2 \longrightarrow HO_2 + CO, \qquad (R12)$$

$$CH_3O + O_2 \longrightarrow H_2CO + HO_2 \qquad (R7)$$

and

$$H + O_2 + M \longrightarrow HO_2 + M. \qquad (R11)$$

The HO_2 radicals may then be recycled back to OH by reaction with NO

$$NO + HO_2 \longrightarrow NO_2 + OH, \qquad (R22)$$

or O_3

$$HO_2 + O_3 \longrightarrow 2O_2 + OH, \qquad (R42)$$

thereby leading to further CH_4 and CO oxidation. Free radicals are eventually destroyed by chain-terminating reactions such as

$$HO_2 + HO_2 \longrightarrow H_2O_2 + O_2, \qquad (R19)$$

followed by

$$H_2O_2 + OH \longrightarrow H_2O + HO_2. \qquad (R21)$$

The methane-oxidation chain can lead to the removal of hydrogen oxides by reactions such as

$$CH_3O_2 + HO_2 \longrightarrow CH_3OOH + O_2, \qquad (R3a)$$

followed by

$$CH_3OOH + OH \longrightarrow CH_3O_2 + H_2O, \qquad (R4b)$$

as well as by heterogeneous loss processes, as discussed later in this section.

The abundance of OH, which determines the rate of photochemical oxidation of CH_4 and CO and the rate of photochemical production of CO, is in part controlled by the abundance of CH_4 and CO, as well as O_3, H_2O and NO, and also the flux of near-UV solar radiation. Therefore, to understand the CH_4–CO photochemical system it is necessary to study a complicated, nonlinear set of photochemical reactions involving hydrogen oxides—OH,

HO_2, H_2O_2; odd nitrogen compounds—NO, NO_2, NO_3, N_2O_5, HNO_2, HNO_3; and odd oxygen compounds—O, $O(^1D)$, O_3. Such studies have often employed a one-dimensional model to analyze the key elements of the tropospheric photochemical system. The models couple the detailed photochemistry of the lower atmosphere with vertical transport, which is parameterized using an eddy-diffusion coefficient. Average eddy-diffusion coefficients for the troposphere are believed to range from about 5×10^4 to about 2×10^5 cm^2/s, with 1×10^5 cm^2/s as a reasonable mean value.[90–92] Representative values for temperature, total number density and relative humidity are generally taken from the U.S. Standard Atmosphere[93] and/or U.S. Standard Atmosphere Supplements.[94] Diurnally averaged photolysis rates are calculated using average sunlight conditions.

While one-dimensional models obviously cannot reproduce the highly variable and turbulent conditions of the troposphere, they have proven to be useful tools in determining the major photochemical processes that control tropospheric abundances of trace gases. For instance, note the qualitative agreement between predicted OH densities of the order of 10^6 mol./cm^3 and initial atmospheric observations of this radical.[95–97] One-dimensional models, when used in conjunction with more detailed two- and three-dimensional models, have also been helpful in estimating the global rate of photochemical production and/or destruction of various trace gases. In particular, the photochemical production and destruction of CH_4 and CO will now be discussed using the one-dimensional model described by Chameides;[98] these calculations are generally consistent with the results of other investigators.[6,14]

To determine the photochemical impact on CH_4 and CO, it is necessary to calculate the profile of OH in the troposphere. Since the photochemical lifetime of OH is only 1 s, OH may be assumed to be in a photochemical steady state, and thus its density is calculated by equating local volume production and destruction rates at each altitude. An iterative technique is employed to ensure consistency between OH and the densities of other species which interact with OH. The resulting OH profile for (diurnally averaged) equinoctial conditions at midlatitudes is illustrated in Figure 8. The NO_x ($NO + NO_2$) profile is illustrated in Figure 9 and is consistent with Noxon's spectroscopic measurements.[99] Recall that the methane-oxidation chain influences the calculated OH abundance, in that the reaction chain converts OH to HO_2 and may also deplete hydrogen oxides. In Figure 10 the net loss of OH and production of HO_2 due to the CH_4–CO oxidation reaction chain is illustrated. The difference between these two rates is the net loss of hydrogen oxides. For comparison, the production rates of OH due to Eqs. (R15) and (R22) are also shown in Figure 10. Thus, CH_4 and CO have a significant effect on the abundance of OH, which leads to complicated

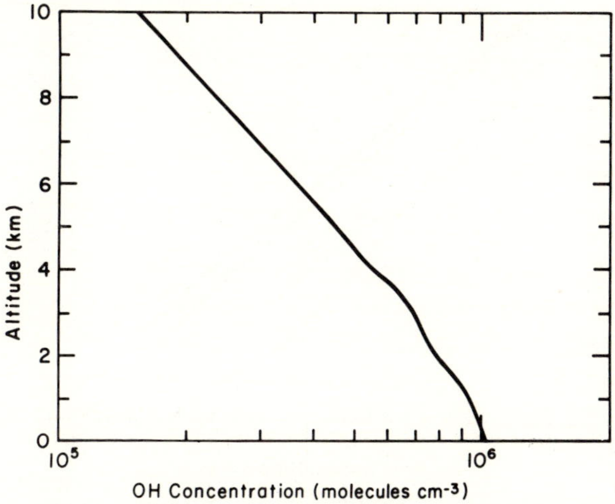

Figure 8 The calculated OH profile.

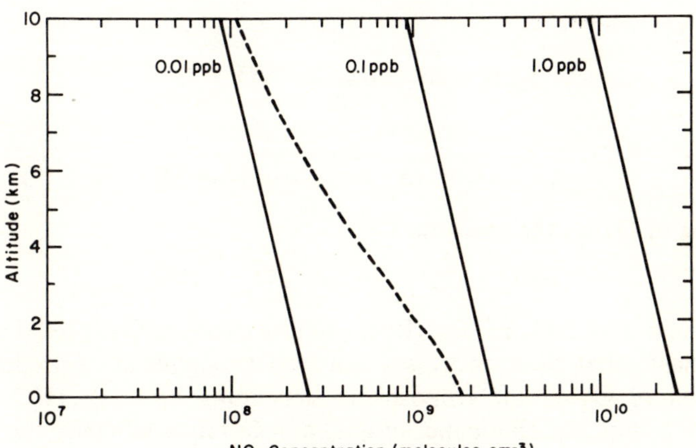

Figure 9 The calculated NO_x ($NO + NO_2$) profile, mol./cm^3. For reference, the heavy continuous curves show approximate lines of constant mixing ratio.[119]

feedback processes between CH_4, CO and OH. The implications of these feedback mechanisms are discussed in Section 8.

The production of HO_2 in the methane-oxidation chain can lead to the local generation of O_3[53] *via* the reaction

$$NO + HO_2 \longrightarrow NO_2 + OH, \qquad (R22)$$

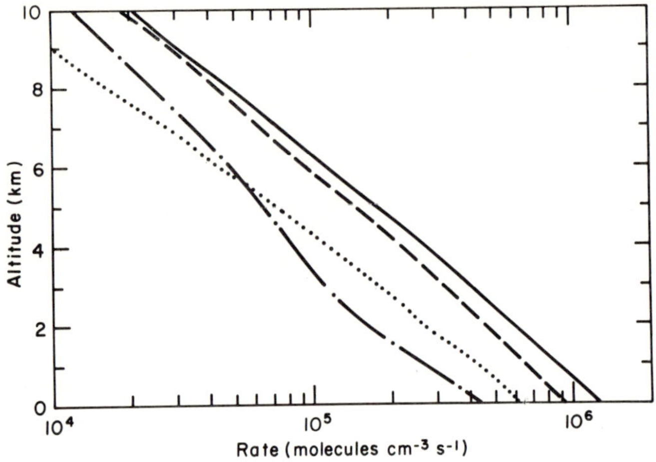

Figure 10 The net rates of OH loss (continuous curve) and HO_2 production (dashed curve) due to the CH_4-oxidation chain. For comparison, the rates of OH production due to Eq. (R15) (dotted curve) and Eq. (R22) (dash–dot curve) are also shown.

followed by

$$NO_2 + h\nu \longrightarrow NO + O \qquad (R30)$$

and

$$O + O_2 + M \longrightarrow O_3 + M, \qquad (R38)$$

or the loss of O_3 *via* the reaction

$$HO_2 + O_3 \longrightarrow 2O_2 + OH. \qquad (R42)$$

The production of NO_2 *via* Eq. (R6c) can also produce O_3. Model calculations indicate that these processes can lead to significant production or destruction of O_3 in the troposphere depending upon the NO_x abundance.[62–64] Thus, the abundance of CH_4 and CO may influence the level of O_3 in the troposphere.

The calculated OH profile in Figure 8 is uncertain, as a result of uncertainties in several key reaction rates and NO_x levels. However, adopting a rate constant of $(3.5 \times 10^{-12})\exp(-1562/T)$ cm^3/mol./s for the reaction of $CH_3CCl_3 + OH \rightarrow CH_2CCl_3 + H_2O$,[65] an atmospheric lifetime of 7 yr is obtained for CH_3CCl_3 when the OH profile in Figure 8 is used. This result is consistent with Singh's inferred CH_3CCl_3 lifetime of 7.2 ± 1.2 yr obtained from the observed CH_3CCl_3 abundance and known anthropogenic emission rates.[100] While the agreement may be fortuitous, it appears that our calculated OH profile approximates global conditions, and the one-dimensional

OH profile may be used to estimate the rates of photochemical destruction of CH_4 and CO. These rates, for an average abundance of 1.4 ppm CH_4 and 0.11 ppm CO, are illustrated in Figure 11 and correspond to integrated column-destruction rates of 5.7×10^{10} CH_4 mol./cm^2/s and 25.6×10^{10} CO mol./cm^2/s. Extrapolating to the Earth's surface yields global loss rates of 2.4×10^{14} g CH_4/yr and 1.9×10^{15} g CO/yr, which imply lifetimes of 15 yr and 0.3 yr for CH_4 and CO, respectively. Note that the CO destruction rate is larger by a factor of 4–5 than the CH_4 loss rate, indicating that even if every CH_4 molecule oxidized by Eq. (R1) is converted to CO, additional CO sources would be needed to balance the large calculated photochemical sink. Furthermore, the CH_4 loss rate is smaller than the estimated CH_4 source in Sections 3 and 4. This result is in diagreement with earlier model calculations, which indicated a closer balance between CH_4 and CO oxidation.[50,67] To a large extent the imbalance presently calculated is due to the revised rate coefficient for Eq. (R13a) used in the calculations. Recent measurements indicate that CO and OH may react through a three-body mechanism at high pressures,[72,101,102] implying that the effective rate constant for Eq. (R13a) in the lower troposphere is about twice as large as was previously believed. This large coefficient for Eq. (R13a) leads to a larger CO loss rate, as well as lower OH densities and therefore smaller CH_4 loss rates.[14] Using the pure two-body rate constant of $(2.1 \times 10^{-13})\exp(-115/T)$ cm^3/mol./s for Eq. (R13a),[74] somewhat more balanced loss rates are obtained for CH_4 and CO, as illustrated in Figure 11. The column-integrated loss rates are

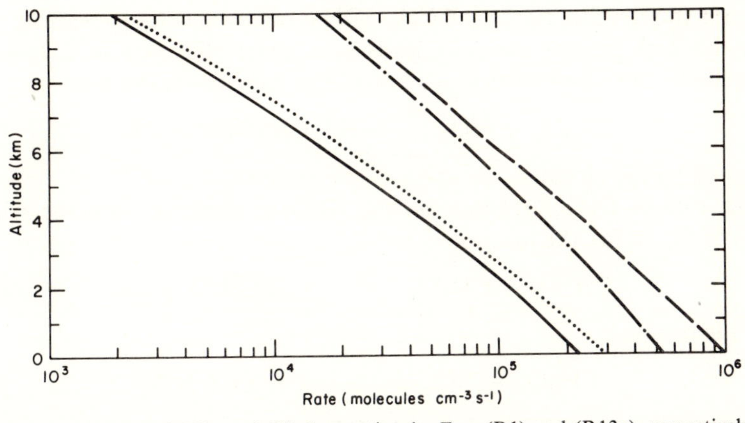

Figure 11 The rates of CH_4 and CO destruction by Eqs. (R1) and (R13a), respectively; the continuous curve is the rate for Eq. (R1) and the dashed curve the rate for Eq. (R13a) with $K_{13a} = (2.1 \times 10^{-13})\exp(-115/T) + (7.3 \times 10^{-33})n_M$ cm^3/mol./s, where n_M is the ambient number density. The dotted curve is the rate for Eq. (R1) and the dash–dot curve is the rate for Eq. (R13a) with $K_{13a} = (2.1 \times 10^{-13})\exp(-115/T)$ cm^3/mol./s.

9.0×10^{10} CH_4 molecules/cm^2/s and 16.4×10^{10} CO mol./cm^2/s, corresponding to global rates of 3.8×10^{14} g CH_4/yr and 1.2×10^{15} g CO/yr.

Inasmuch as the mechanism for the three-body reaction of CO and OH is not understood, it is not possible to rule out the possibility that the larger measured rate for Eq. (R13a) at high pressures was due to contamination. Thus, the rates of CH_4 and CO destruction are uncertain, and therefore the global destruction rates may be estimated at (2–4) $\times 10^{14}$ g CH_4/yr and (1.2–1.9) $\times 10^{15}$ g CO/yr.

To determine the rate of production of CO from the oxidation of CH_4, it is necessary to determine the efficiency of CH_4 conversion to CO. Assuming the basic sequence of homogeneous gas-phase reactions in Figure 7, the conversion efficiency would be unity in the absence of any heterogeneous removal. However, if heterogeneous processes tend to remove a significant fraction of an intermediate species in the methane-oxidation chain, the conversion efficiency is less than unity. For heterogeneous removal to constitute a significant sink for an intermediate species, that species probably must have a high solubility and a photochemical lifetime comparable to or greater than about 10 days, the characteristic time for heterogeneous removal in the lower atmosphere.[103] Since calculations yield a CH_3OOH photochemical lifetime of about 8 days near the ground to about 12 days at 10 km, it is likely that heterogeneous loss of CH_3OOH is significant. (By comparison, the H_2CO photochemical lifetime is only one-third of a day, and the other intermediates in the methane-oxidation chain have even shorter lifetimes.) Thus, a rainout or heterogeneous loss reaction for CH_3OOH—Eq. (R46)—has been included in the reaction scheme, at a first-order rate of $1 \times 10^{-6}\ s^{-1}$. Inclusion of this reaction implies that 50% of all CH_3OOH is removed heterogeneously. Note that this loss process not only leads to a lower efficiency of CH_4-to-CO conversion, but also constitutes a loss of free radicals *via* the reaction

$$CH_3O_2 + HO_2 \longrightarrow CH_3OOH + O_2, \qquad \text{(R3a)}$$

followed by Eq. (R4b).

The rate of CH_3OOH loss *via* Eq. (R4b) is dependent on the relative rates of Eq. (R3a), the reaction

$$CH_3O_2 + CH_3O_2 \longrightarrow 2CH_3O + O_2 \qquad \text{(R3b)}$$

and the reaction

$$CH_3O_2 + NO \longrightarrow CH_3O + NO_2, \qquad \text{(R6c)}$$

since Eqs. (R3b) and (R6c) bypass CH_3OOH and thus preclude the possibility of heterogeneous removal. Figure 12 illustrates the branching ratios for Eqs. (R3a), (R3b) and (R6c), indicating that about 80% of the CH_3O_2 reacts to produce CH_3OOH. Since about 50% of the CH_3OOH is removed hetero-

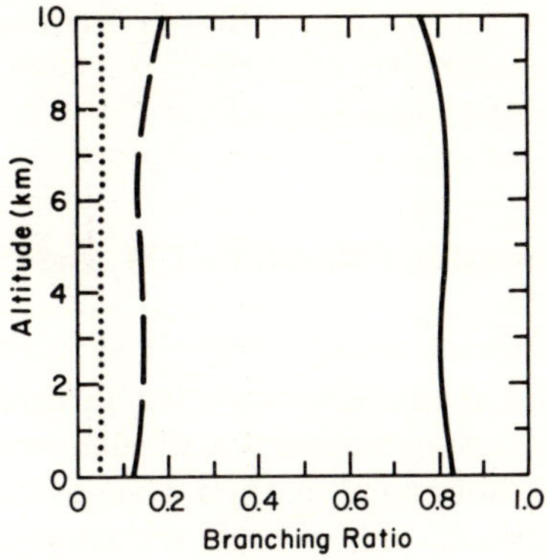

Figure 12 Branching ratios for CH_3O_2 reactions. The continuous curve represents the probability that a CH_3O_2 radical will undergo Eq. (R3a), the dashed curve corresponds to Eq. (R6c) and the dotted curve to Eq. (R3a).

geneously, it may be inferred that about 60% of the CH_4 oxidized in Eq. (R1) is eventually converted to CO. Detailed model calculations confirm this estimate for the efficiency of CH_4-to-CO conversion. However, this factor is highly uncertain. Note in Table VI that almost all the rate constants of importance in determining the relative rate of heterogeneous CH_3OOH loss have been estimated rather than directly measured. Thus, the conversion efficiency is subject to modification as the relevant rate coefficients are measured. The conversion efficiency is also dependent on the NO_x density since an increase in NO would enhance the rate of Eq. (R6c) relative to Eq. (R3a), leading to less CH_3OOH production and less heterogeneous loss.[104] For instance, increasing the NO_x density by a factor of 5 increases the conversion efficiency to about 70%. Given the uncertainties in rate constants and NO_x levels, the conversion efficiency may be estimated to lie in the range 20–100%, with 60% as the "best estimate." The CH_4 destruction rate yields a CO production rate of (1–9) $\times 10^{10}$ CO mol./cm^2/s or, globally, 7×10^{13} to 6×10^{14} g CO/yr.

It is also possible that CO is photochemically generated from the degradation of nonmethane hydrocarbons (NMHC) in the atmosphere. Highly reactive NMHC are produced anthropogenically—primarily as a result of incomplete combustion—at a rate of about 2.9×10^{13} g/yr (1968 estimate).[105] Thus, the photochemical degradation of anthropogenically produced hydrocarbons may provide only a small source of CO. However, it is possible that less-reactive NMHC, such as C_2H_6, C_2H_2, C_3H_6, C_4H_{10} and C_5H_{12}, are present in the atmosphere at levels approaching 0.1–1.0 ppb,[106–109] and could lead to significant CO production as a result of reaction paths similar to the methane-oxidation chain. Chameides and

Cicerone estimate that the CO source due to NHMC may be as much as one-third of the source from CH_4 oxidation.[110] The production of CO from NMHC may therefore be estimated at about 2×10^{13} to 2×10^{14} g/yr.

7 Two- and Three-dimensional Models for CH_4 and CO

The one-dimensional models generally rely on eddy-diffusion coefficients for vertical transport and make no allowance for horizontal transport. Newell recently questioned the results obtained with such models since they ignore large-scale convective horizontal motion transporting material between vertical columns.[19] Certainly if horizontal transport is important, then the two- and three-dimensionality of these processes must be considered. For species with relatively short residence times, such considerations indeed become necessary.

Horizontal transport is of the utmost importance in understanding the impact of possible interhemispheric exchange on concentration differences between the Northern and Southern Hemispheres. Czeplak and Junge have approximated the troposphere with respect to this phenomenon by a two-box model, and analyzed much of the available tracer data on this basis.[111] Their results strongly suggested an interhemispheric exchange time of close to a year. As a result, a species with a residence time of less than a year could show distinct asymmetry between the Northern and Southern Hemispheres if the source and/or sink distribution of the species is substantially different between the two Hemispheres. Newell *et al.* employed the meridional component of the wind in the equatorial-pressure–longitude plane to estimate the seasonal transport of CO from the Northern to the Southern Hemisphere.[48] This interhemispheric transport of CO ranged from 1.2×10^{14} g/yr for December–February to 2.3×10^{14} g/yr for June–August, with an annual average of 1.6×10^{14} g/yr. This rate of exchange can be compared with anthropogenic emissions of CO of somewhere between 2×10^{14} and 6×10^{14} g/yr. These calculations would thus indicate that interhemispheric exchange could restrict the distribution of much of the anthropogenic CO emissions to the Northern Hemisphere.[48] As a result of these considerations, two- or three-dimensional models are needed to elucidate some of these factors more completely.

Wofsy developed a two-dimensional (altitude–latitude) model to investigate the global budgets of CH_4 and CO.[6] His study was based on the postulate that the OH-radical destruction rate is primarily due to reaction with CH_4 and CO, and the formation rate of OH can be predicted on the basis of known values of the temperature, relative humidity, O_3 concentration and solar insolation. Thus, he used known CO and CH_4 distributions (actually these distributions were based on very sparse data), along with distributions of the other parameters and a detailed reaction-chemistry model, to calculate a two-dimensional OH distribution. Photodissociation rates were averaged over a 24-h period appropriate to the latitude, and the results were presented for solar declination angles of 23° and −23°.

Wofsy's calculations predicted 30–40% more OH at southern midlatitudes than at corresponding northern locations, which he attributed mainly to the asymmetry of the CO distribution. He further concluded that oxidation of CH_4 and anthropogenic production are the major sources of atmospheric CO. The chemical source of CO was found to be nearly an order of magnitude greater than all other CO sources in the Southern Hemisphere, but natural and anthropogenic sources were comparable in the Northern Hemisphere. On the basis of the model, Wofsy stated that the annual mean lifetime for CO is less than three months, but the CH_4 lifetime is greater than two years.

More recently, Crutzen and Fishman carried out similar two-dimensional model calculations for the Northern Hemisphere.[14] In these calculations, the faster rate coefficient for the reaction CO + OH implied by recent laboratory experiments (see Section 6) was employed. The new rate coefficient led to lower OH levels and, as a result, smaller production of CO from CH_4 oxidation.

These models do not predict the concentration of CH_4 or CO.[6,14] Kwok *et al.* reported a three-dimensional simulation for CO including two vertical levels which was predictive, although the formation of CO by CH_4 oxidation was disregarded.[112] A general circulation model was used to develop the simulated wind field, and advection was the only mechanism for horizontal CO transport, transport by subgrid-scale eddies being neglected. Vertical transport occurred by large-scale advection, and numerical experiments in which vertical redistribution by cumulus convection was taken into account were also reported. It was assumed that CO is chemically inert, and only anthropogenic sources of CO were considered.

The general circulation model was run for 24 simulated days to build up a realistic circulation pattern, and then the CO sources began operating. Initially there was no CO present in the atmosphere. The CO isopleths associated with sources from the United States extended well out over the northern Atlantic Ocean after only two days. Since these sources are located in a strong belt of prevailing westerly winds, the isopleths connected to those from European sources; this occurred after about six days. The results are shown in Figure 13. The simulation continued for 28 days from the time that the sources began operating, and this chaining of the CO isopleths became more intense. As can be seen, there was essentially no transport of CO into the Southern Hemisphere during this period.

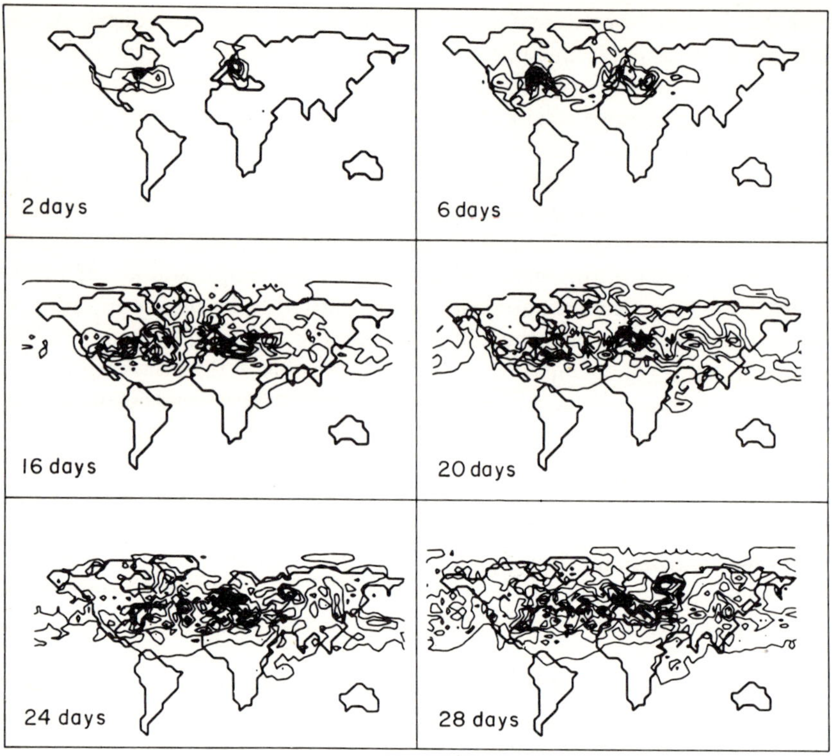

Figure 13 CO-concentration contours in the lower layer following the inclusion of sources in the model; the contour interval is 10 ppb.[112]

More recently, Peters and Jouvanis developed a three-dimensional model that incorporated the chemistry of CH_4 and CO, surface–air exchange, and advection–diffusion transport.[113] The troposphere was divided into five levels. The chemistry component of the model used to describe the atmospheric transformation of CH_4 to CO consisted of 20 reactions; the pseudosteady-state approximation was used for all intermediate species. The coupled partial differential equations for CH_4 and CO were numerically solved.

The general approach consisted of the following steps.

a) The CH_4 and CO concentrations in the troposphere were initialized at 1.5 ppm CH_4 and 0.1 ppm CO—approximately the global average values.

b) The sources and sinks of the various species were distributed over the Earth's surface in accordance with physicochemical considerations; this involved the proper interpretation of ocean and land as sources and/or sinks of the particular species. Leakage to the stratosphere was simulated using an exchange coefficient with a concentration driving force dependent on the upper troposphere concentration.

c) The coupled unsteady-state turbulent diffusion equations for CH_4 and CO, with the boundary conditions established by (b), were numerically integrated.

The inherent advantage of this procedure is that the final atmospheric concentrations of the pollutant species being studied are not presupposed.

The basis for the model parameters was January climatological data, which included the horizontal wind field, temperature and dew point. A vertical velocity was obtained from the horizontal wind field using the continuity equation. The chemical-reaction rate constants were based on the temperature field for those constants that showed a substantial temperature dependence. The chemical-reaction model also required water vapor, O_3, NO and NO_2 concentrations as input. The water-vapor concentration was established from the dew-point data where possible.

The land and ocean areas were differentiated according to pollution-source strength functions and their capacity to exchange the gaseous pollutants with the air. Since the oceans apparently act as sources of both CH_4 and CO, those locations assumed a supersaturation so as to describe the source. For methane, supersaturation corresponded to 1.8 ppm for ocean areas and 1.0 ppm for land areas, whereas for carbon monoxide these values were 3.5 ppm and 0.2 ppm, respectively. In those grids that had both ocean and land areas, a weighted average corresponding to the fraction of ocean and land area was used. In view of the limited knowledge of species interaction with the various soil types, there was no differentiation according to the type of land area.

It was necessary to consider the position of the sun to account for the photochemical-reaction rate constants and the apparent diurnal nature of the supersaturation of CO in the oceans. The declination angle was set at $-20°$, consistent with about mid-January, and the initial time for the integration corresponded to 12:00 Greenwich mean time. Two computer runs, each for approximately 20-day simulations, were reported: a simulation of the base model, and a simulation in which the anthropogenic sources of CO were made five times as large.

The results obtained again exhibited the isopleth chaining.[113] This is shown in Figure 14, where the ground-level concentrations of CO are plotted as isopleths. The high-concentration regions over the United States, the western North Atlantic Ocean, Europe and Japan reportedly resulted from the substantial anthropogenic sources in these regions. The highest-shown ground-level CO concentration was 0.203 ppm, located just off the eastern

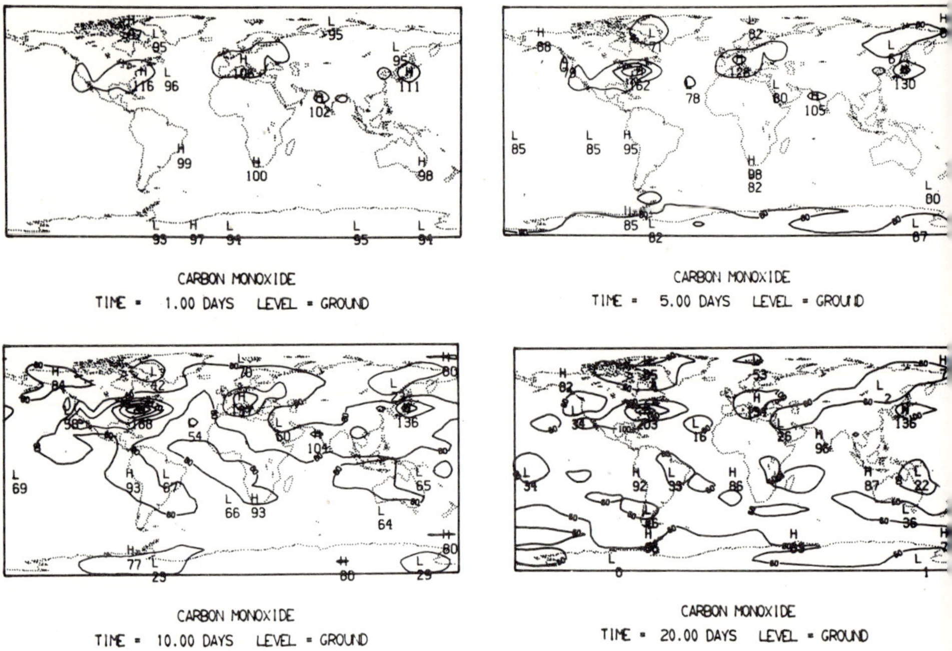

Figure 14 Simulation results with present-day anthropogenic sources of CO; the concentrations on the isopleths are in ppb.[113]

United States coast. This value is of the same order as many of the measurements made over the Atlantic Ocean summarized by Seiler.[5]

It is important to note that high CO concentrations were not predicted over the Pacific Ocean. (This is not inconsistent with Seiler's observations—Figures 3 and 4—since these were heavily weighted by measurements over the Atlantic Ocean.[5]) The results would be consistent with a relatively short residence time of CO since the anthropogenic emissions would have been substantially consumed by reaction with OH by the time they arrived (moving in an easterly direction) over the Pacific. Substantial longitudinal variation of the CO concentrations was indicated.[113] Furthermore, the maximum ground-level CO concentrations were predicted to be in the region of 30–55°N. Large-scale transport meant that these regions were downwind of the major anthropogenic sources. This large-scale transport, in combination with the reduced OH available for CO consumption at the higher latitudes, is probably responsible for Seiler's observation that the maximum CO concentration appears in the region 60–80°N.[5]

In Figure 15 similar results of the simulation with the increased anthropogenic sources are shown. The effect of increasing the anthropogenic CO

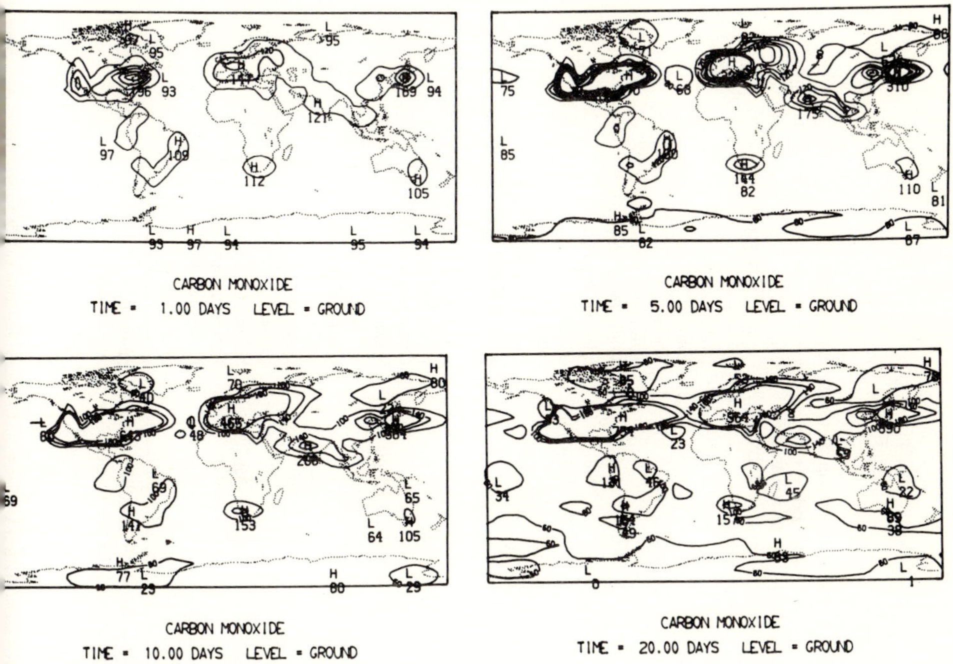

Figure 15 Simulation results with five times the present-day anthropogenic sources of CO; the concentrations on the isopleths are in ppb.[113]

emissions by a factor of five was quite pronounced. The peak CO concentrations were in the same geographic regions and increased by approximately a factor of four. It was suggested that the reaction of CO with OH, which is a strong sink of CO, was responsible for some suppression of the increase. It is also apparent that the concentrations in the Southern Hemisphere have not been substantially affected, which would be anticipated with simulations of this duration, since interhemispheric transport is slow in comparison with the time of the simulation.

Peters and Jouvanis also compared the results of the first simulation with the CH_4 and CO budgets suggested by other investigators on other bases; the results are summarized in Table VII.[113] Among other things, the data show that interhemispheric exchange greatly restricts the CO transport to the Southern Hemisphere. This restriction and the concentration of anthropogenic CO sources in the Northern Hemisphere seem largely responsible for the asymmetry of the observed CO concentrations. The multidimensional models are consistent in that regard, and point up the value of using such models for trace species that have reasonably short residence times in the atmosphere.

TABLE VII

Comparison of predicted budgets and parameters with other studies (from Peters and Jouvanis[113])

Feature	Simulation result (10^{14} g/yr)	Other studies (10^{14} g/yr)	Reference
Natural CH_4 source[a]	10	9	McConnell *et al.*[51]
		5.5–11	Ehhalt[4]
		7.3	Wofsy[6]
Anthropogenic CO source	3.2	2.0	Kummler and Baurer[114]
		2.0	Weinstock and Niki[115]
		5.8	Chameides *et al.*[20]
		6.4	Seiler[5]
		12[b]	Wofsy[6]
Oceanic source of CH_4	0.015	0.047–0.21	Ehhalt[4]
Oceanic source of CO	0.083	0.15	Lamontagne *et al.*[24]
		1.0	Seiler[5]
		1.6	Junge *et al.*[34]
		2.2	Linnenbom *et al.*[25]
CO source by CH_4 oxidation	19	7	Chameides *et al.*[20]
		15–40	Seiler[5]
		13[b]	Wofsy[6]
Oxidation of CO to CO_2	24	15	Chameides *et al.*[20]
		19–50	Seiler[5]
		20[b]	Wofsy[6]
Leakage of CH_4 to stratosphere	4.7	1–2	Ehhalt[4]
Leakage of CO to stratosphere	0.18	1.1	Seiler[5]
		0.6[c]	Pressman and Warneck[49]
		0.1	Newell *et al.*[48]
Interhemispheric exchange of CO (north to south)	1.2	1.6	Newell *et al.*[48]

[a] Includes the anthropogenic sources of CH_4, which represent a relatively small fraction of the total.
[b] Represents 90% of the values listed, since that reportedly accounted for the tropospheric contribution.
[c] Represents 11% of the CO inventory.

8 Possible Future Variations in the Atmospheric Abundance of CH_4 and CO

The possibility of secular variations in the global abundance of CH_4 and CO is cause for concern, since these variations may lead to perturbations in other key environment parameters. In Section 6, we noted that CH_4 and CO play a central role in the tropospheric photochemistry and thus affect the concentration of OH and other free radicals. Variations in tropospheric OH could lead to perturbations in many other trace gases which are scavenged by OH, such as CH_3Cl, CH_3CCl_3, CHF_2Cl, $CHFCl_2$, SO_2, NH_3, NO_2 and non-methane hydrocarbons. In the case of chlorine-containing compounds, these variations could lead to perturbations in the stratospheric chlorine abundance, and thus to changes in the O_3 concentration in the stratosphere.[55] Furthermore, as discussed in Section 5, CH_4 and CO variations can directly affect stratospheric O_3. As a result of uncertainties in several key reaction rates and the difficulties involved in coupling the tropospheric and stratospheric photochemistry, the net effect of a change in the tropospheric CO and CH_4 abundances on stratospheric O_3 is still unclear, and further investigation is needed. (The implications of a perturbation to stratospheric O_3 include climatic variations and increases in the incidence of skin cancer.[116])

In addition to variations in stratospheric O_3, changes in CH_4 may lead directly to climatic perturbations. Radiative-transfer calculations indicate that the 7.6-μm absorption band of CH_4 plays a significant role in the atmosphere's radiative balance and that a doubling of the CH_4 abundance may cause an increase in the Earth's surface temperature by as much as 0.3 K as a result of the "greenhouse effect."[117] Since CH_4 produces H_2O in the stratosphere, and the increase in surface temperature corresponding to a doubling of the stratospheric H_2O abundance has been estimated at 1.03 K,[117] the effects of a CH_4 increase may be even larger.

In view of the apparent importance of CH_4 and CO, it is of interest to identify any possible causes of future CH_4 and CO perturbations and to estimate the magnitude of these variations. In this regard, it is important to note that the CH_4 and CO photochemistry is highly coupled, and a variation in one is likely to cause a variation in the other. Since both CH_4 and CO are consumed by reactions with OH (see Section 6),

$$CH_4 + OH \longrightarrow CH_3 + H_2O, \tag{R1}$$

$$CO + OH \longrightarrow CO_2 + H, \tag{R13a}$$

an increase or decrease in tropospheric OH would tend to cause a decrease or increase, respectively, in CH_4 and CO concentrations. However, the abundance of OH is determined by the concentration of CO and CH_4 since, as indicated in Table VIII, Eqs. (R1) and (R13a) represent the major loss paths for OH. This coupling implies a positive feedback; i.e., a perturbation in

TABLE VIII

Calculated OH production and destruction rates (10^5 mol./cm^3/s) near the ground for equinoctial conditions at midlatitudes

	OH source	Rate		OH sink	Rate
(R15)	$O(^1D) + H_2O$	6.2	(R13a)	$CO + OH$	9.1
(R22)	$HO_2 + NO$	4.5	(R1)	$CH_4 + OH$	2.2
(R42)	$HO_2 + O_3$	2.5	(R21)	$H_2O_2 + OH$	1.0
(R20a)	$H_2O_2 + h\nu$	1.6	(R14)	$H_2 + OH$	0.8
(R5)	$CH_3OOH + h\nu$	0.6	(R4a)	$CH_3OOH + OH$	0.8
			(R8)	$H_2CO + OH$	0.4
	Total	15	(R23)	$NO_2 + OH$	0.2
			(R17)	$HO_2 + OH$	0.1
				Total	15

either CO, CH_4 or OH will be amplified as a result of subsequent variations in the other species. For instance, if an increased source of CO led to a higher CO density, OH concentrations would decrease, thereby leading to further increases in CO, as well as increases in CH_4. This positive-feedback system is offset to some extent by the interaction of OH and NO_x. Note in Table VIII that NO_x ($NO + NO_2$) tends to enhance OH *via* the reaction

$$NO + HO_2 \longrightarrow NO_2 + OH. \qquad \text{(R22)}$$

Since the major sink of NO_x is conversion to HNO_3 by reaction with OH,

$$NO_2 + OH \xrightarrow{M} HNO_3, \qquad \text{(R23)}$$

the NO_x–OH coupling acts to stabilize OH levels.

Studies using photochemical models of the troposphere have indicated that significant perturbations in OH, CO, and CH_4 may result from the continued anthropogenic production of CO and NO_x.[20,118–120] The anthropogenic production of CO and NO_x occurs primarily in combustion-related processes; CO emissions are mainly associated with mobile sources, while for NO_x both mobile and stationary sources are important. These emissions grew at a rate of about 4.5%/yr for CO[121] and 5%/yr for NO_x[122] before 1967, but the introduction of pollution-control devices on automobiles sold in the United States after 1967 has slowed the rate of anthropogenic CO and NO_x emissions there. According to data compiled by the Environmental Protection Agency in 1970–74,[47] anthropogenic CO emissions in the United States have decreased by about 12%, while NO_x emissions have increased by only 10%. The global emission trends over the past decade and for future decades, however, are uncertain. It is probable that, with continued industrial expansion outside the United States, the global anthropogenic CO and NO_x

emission rates will continue to grow in the coming decades, although at a slower rate than before the introduction of pollution-control equipment.

Chameides, who estimated that presently about one-third of the CO and NO_x in the atmosphere is of anthropogenic origin, found that increased emissions of CO and NO_x may lead to significant perturbations in the CH_4–CO–OH system by the year 2000.[119] Since CO tends to decrease the OH concentration, while NO_x enhances the OH concentration, simultaneous increases in CO and NO_x tend to oppose each other, and the direction of the perturbation depends on the relative increases in CO and NO_x. Illustrative calculations by Chameides for various emission rate patterns are shown in Figures 16 and 17.[119] These graphs indicate that, depending on the rate of future anthropogenic emissions, increases or decreases in CH_4 of as much as 40% may occur in the coming decades. It is also possible that the CO and NO_x emissions will increase at rates which will lead to a cancellation of effects and a negligible perturbation; in Figures 16 and 17 this case most closely corresponds to curves IV, in which NO_x emissions grow at twice the rate of CO emissions.

While these results are uncertain and should be taken as an approximate indication of possible future trends, it does appear that significant perturbations in the tropospheric photochemical system may occur as a result of emissions of CO and NO_x. In view of this possibility, long-term monitoring programs are needed to determine the detailed global distributions of CO and NO_x, as well as OH and CH_4. Continued laboratory experiments are also needed to permit a thorough characterization of the tropospheric photochemical system. Finally, the further development and implementation

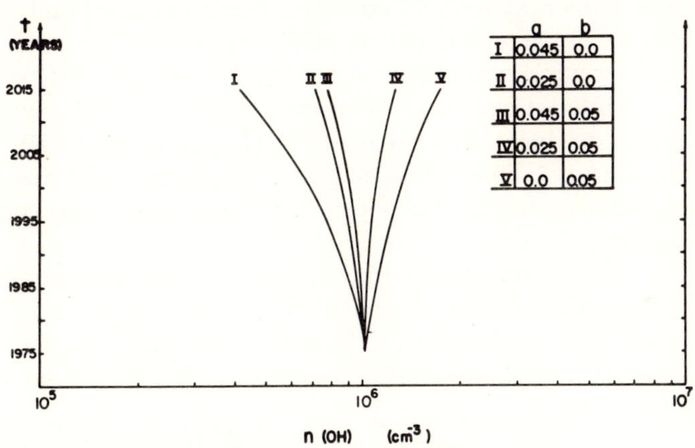

Figure 16 Time variation in ground-level OH concentrations for various patterns of CO and NO_x emission rate: (a) yearly rate of growth in the CO anthropogenic emission rate; (b) the same for NO_x.[119]

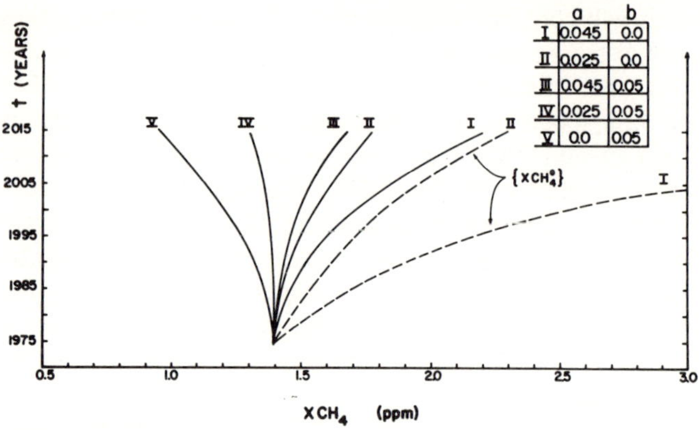

Figure 17 Time variation of the CH_4 mixing ratio (continuous curves) and the equilibrium CH_4 mixing ratio $x^{\circ}_{CH_4}$ (dashed curves), for various emission-rate patterns: (a) yearly rate of growth in the CO anthropogenic emission rate; (b) the same for NO_x.[119]

of pollution-control devices for combustion systems, as well as the use of alternative power sources, may alleviate future perturbations to the tropospheric photochemistry.

The OH concentration is also sensitive to the CH_4 abundance, although to a lesser extent than to CO, and variations in the sources of CH_4 can lead to perturbations in the tropospheric photochemistry, as well as in the CH_4 abundance. The production of CH_4 is primarily the result of separate processes occurring in anaerobic environments, as discussed in previous sections. However, it is not known whether the CH_4 source is presently increasing, decreasing or remaining roughly constant. While increased cultivation of rice paddy fields would tend to enhance CH_4 production, the destruction of wetlands and swamp environments would lead to a smaller CH_4 source. In this regard, it is interesting to note an estimate that the wetlands region of the Sudd in Sudan produces about 6×10^{13} g CH_4/yr or about 10% of the global CH_4 source.[38] Thus, disturbing the ecosystem of the Sudd, as an example, could have long-range implications for atmospheric photochemistry and the climate.

Variations in CH_4 and CO due to changes in the OH concentration may also result from perturbations in the abundance of stratospheric O_3. The major source of OH involves the production of metastable oxygen atoms in the photolysis of O_3 by photons with wavelengths of less than about 3100 Å[59]

$$O_3 + h\nu(\mathrm{UV}) \longrightarrow O(^1D) + O_2, \qquad \text{(R39)}$$

followed by

$$H_2O + O(^1D) \longrightarrow 2OH. \qquad \text{(R15)}$$

It has been pointed out that a perturbation in stratospheric O_3 can lead to variations in the $O(^1D)$ production rate and thus in OH, CH_4 and CO levels in two ways: (a) by changing the flux of near-UV radiation penetrating the tropopause; and (b) by changing the O_3 density in the lower stratosphere, and thus the injection rate of stratospheric O_3 into the troposphere.[61,20] Table IX illustrates the calculated dependence of $x^{\circ}_{CH_4}$, the steady-state CH_4 mixing ratio, on the stratospheric O_3 abundance, $N(O_3)$, and the O_3 mixing ratio at 11 km. (These calculations are similar to those of Chameides *et al.*,[20] but include recent revisions to key reaction-rate coefficients.) Since model calculations of the stratosphere predict that a decrease in $N(O_3)$ is generally accompanied by an increase in the O_3 density near the tropopause, these results indicate that significant CH_4 (as well as CO, CH_3Cl, etc.) perturbations may occur from a change in stratospheric O_3. For instance, if a 10% decrease in $N(O_3)$ is accompanied by a 25% increase in x_{O_3} at 11 km, an approximate 30% decrease in $x^{\circ}_{CH_4}$ is predicted.

TABLE IX

The equilibrium CH_4 mixing ratio, $x^{\circ}_{CH_4}$ (ppm), as a function of stratospheric O_3 column, $N(O_3)$, and O_3 mixing ratio at 11 km (x_{O_3} at 11 km)

	$N(O_3)$(mol./cm²)				
x_{O_3} at 11 km (ppm)	7×10^{18}	8×10^{18}	9×10^{18}	10×10^{18}	11×10^{18}
0.065	0.89	1.2	1.7	2.3	3.3
0.095	0.81	1.1	1.6	2.2	3.2
0.13	0.72	1.0	1.5[a]	2.1	3.0
0.165	0.64	0.96	1.3	1.9	2.7
0.20	0.56	0.87	1.2	1.7	2.4

[a] Present-day conditions assumed to be $N(O_3) = 9 \times 10^{18}$ mol./cm², x_{O_3}(11 km) = 0.13 ppm and $x^{\circ}_{CH_4} = 1.5$ ppm.

Since the stratospheric O_3 abundance is influenced by the level of CH_4 and CO in the atmosphere, and the CH_4 and CO levels are dependent on the stratospheric O_3 abundance, there are feedback processes which couple the tropospheric and stratospheric photochemistries.[20,61] Whether these feedback processes tend to stabilize or destabilize the tropospheric and stratospheric photochemistry depends on whether the net effect of an increase in CH_4 and CO is to destroy or enhance stratospheric O_3. While increases in stratospheric CH_4 and CO from increases in the tropospheric CH_4 and CO abundance probably lead to a stratospheric O_3 increase, increases in tropospheric CH_4 and CO also tend to lower tropospheric OH, thereby enhancing the concentrations of CH_3Cl and CH_3CCl_3, which tends to decrease stratospheric O_3. At present, it is not known which of these effects will dominate, and further modeling and laboratory experiments are needed to determine the direction and magnitude of the tropospheric–stratospheric photochemical-feedback system.

Thus, calculations indicate that perturbations in atmospheric CH_4 and CO levels of as much as 30–40% may occur in the coming decades. The consequences of a perturbation in CH_4 and CO may include variations in stratospheric O_3 and climatic changes. However, it should be pointed out that the rather limited record of CO and CH_4 measurements compiled to date does not indicate any discernible secular trends in CO and CH_4 over the past decade. Nevertheless, we believe that a concerted and long-term monitoring program for CO and CH_4 should be undertaken to determine whether any variations occur in the future.

9 Conclusions

It is appropriate to conclude this paper by summarizing what we believe to be the best present-day estimates of the CH_4 and CO global budgets; these are presented in Tables X and XI. Recognizing the difficulties in assigning a specific rate to a global process, we have used ranges of rates for each production or destruction process. Nevertheless, it is possible that some of the rates in Tables X and XI are in error, and may have to be revised as more is learnt about the dynamics and chemistry of the troposphere and new data become available. Given that the atmosphere contains about 4.4×10^{15} g CH_4 and 5.6×10^{14} g CO, we infer atmospheric residence times based on the source strengths of 5–9 yr for CH_4 and 0.3–0.8 yr for CO.

One of the more striking aspects of the CO budget is the large role of anthropogenic production. Our estimates indicate that man-related processes produce at least 30% of the CO found in the troposphere, in agreement with

TABLE X

Global budget of tropospheric CH_4

Sources	Strength (10^{14} g/yr)
Oceans	0.04–0.2
Soils	5–8
Anthropogenic	0.1–0.5
Total	5–9
Sinks	**Strength (10^{14} g/yr)**
CH_4 + OH	2–4
Transport to stratosphere	0.1–0.7
Total	2–5

TABLE XI

Global budget of tropospheric CO

Sources	Strength (10^{14} g/yr)
Oceans	0.6–3
Anthropogenic	5–7
CH_4 + OH	0.7–6
NMHC oxidation	0.2–2
Total	7–18
Sinks	**Strength (10^{14} g/yr)**
Soils	3–5
CO + OH	12–19
Transport to stratosphere	0.1–0.6
Total	15–25

Seiler's earlier calculations.[5] Given that most anthropogenic CO is produced in the Northern Hemisphere, it is therefore likely that the hemispheric asymmetry in CO (see Figures 2 and 3) is due to the large production of CO by man. Since increases in CO may lead to serious perturbations in the atmosphere (see Section 8), it is imperative that we begin a long-term monitoring program to determine if CO levels are on the rise globally.

The major sink of both CO and CH_4 is photochemical destruction *via* reaction with OH, further emphasizing the coupling between CO and CH_4 abundance and the tropospheric photochemistry, as discussed in Section 6. Our study, in agreement with previous calculations, indicates that soils may represent a significant sink of atmospheric CO, while for CH_4 soils represent the major source to the atmosphere. Thus, the interaction between the surface and atmosphere of the Earth plays a central role in regulating the CH_4–CO system. Since CH_4 and CO exert some influence on the climate, it is possible that the biosphere, by controlling production and destruction rates of CH_4 and CO at the soil–air interface, acts to stabilize the atmospheric environment. This interesting possibility of a biospheric–atmospheric feedback system is similar to mechanisms proposed in the Gaia hypothesis[123]; more interdisciplinary research involving the biological, chemical and physical sciences is needed in order to determine whether such a feedback system exists and how it operates.

The budgets of CH_4 and CO, as presented in Tables X and XI, appear to be in approximate balance, although CH_4 sources are larger than sinks, while CO sinks tend to outweigh sources. These small imbalances could indicate the existence of another, as yet unidentified, sink of CH_4 or source of CO, or may be indicative of the need for adjustments to the rates of processes already included in the tables. For instance, the balance in the budgets could be improved by increasing the loss of CH_4 due to reaction with OH, which would also increase the production rate of CO. However, while an increase in OH would lead to a larger rate for the CH_4 + OH reaction, it would also cause a faster loss of CO through the reaction of CO + OH. In conclusion, while the budgets of tropospheric CH_4 and CO are in approximate balance, we believe that further research is needed to improve and refine our understanding of the global CH_4 and CO budgets. Measurement programs to monitor atmospheric levels of CH_4 and CO should be started, so that any future variations in their abundances may be identified.

ACKNOWLEDGEMENTS

This work was supported by NASA Research Grants NSG 1424 and NSG 1497.

REFERENCES

1. M. V. Migeotte, "Spectroscopic evidence of methane in the Earth's atmosphere," *Phys. Rev.*, **73**, 519 (1948).
2. M. V. Migeotte, "The fundamental bond of carbon monoxide at 4.7 μ in the solar spectrum," *Phys. Rev.*, **75**, 1108 (1949).
3. D. R. Bates and A. E. Witherspoon, "The photochemistry of some minor constituents of the Earth's atmosphere (CO_2, CO, CH_4, N_2O)," *Mon. Not. R. Astron. Soc.*, **112**, 101 (1952).
4. D. H. Ehhalt, "The atmospheric cycle of methane," *Tellus*, **26**, 58 (1974).
5. W. Seiler, "The cycle of atmospheric CO," *Tellus*, **26**, 116 (1974).
6. S. C. Wofsy, "Interactions of CH_4 and CO in the Earth's atmosphere," *Ann. Rev. Earth Planet, Sci.*, **4**, 441 (1976).
7. L. S. Jaffe, "Carbon monoxide in the biosphere: sources, distribution and concentrations," *J. Geophys. Res.*, **78**, 5293 (1973).
8. P. Urone, "The primary air pollutants—gaseous; their occurrence, sources and effects," in *Air Pollution*, Vol. 1, edited by A. C. Stern (Academic, New York, 1976) pp. 23–75.
9. M. H. Bortner, R. H. Kummler and L. S. Jaffe, *A Review of Carbon Monoxide Sources, Sinks and Concentrations in the Earth's Atmosphere*, NASA CR-2081 (1972).
10. B. D. Tebbens, "Gaseous pollutants in the air," in *Air Pollution*, Vol. 1, edited by A. C. Stern (Academic, New York, 1968) pp. 23–46.
11. W. Seiler and C. Junge, "Carbon monoxide in the atmosphere," *J. Geophys. Res.*, **75**, 2217 (1970).
12. D. H. Ehhalt and L. E. Heidt, "Vertical profiles of CH_4 in the troposphere and stratosphere," *J. Geophys. Res.*, **78**, 5265 (1973).
13. P. E. Wilkniss, R. A. Lamontagne, R. E. Larson, J. W. Swinnerton, C. R. Dickson and T. Thompson, "Atmospheric trace gases in the Southern Hemisphere," *Nature Physical Science*, **245**, 45 (1973).
14. P. J. Crutzen and J. Fishman, "Average concentrations of OH in the Northern Hemisphere troposphere and the budgets of CH_4, CO and H_2," *Geophys. Res. Lett.*, **4**, 321 (1977).
15. C. E. Junge, "Residence time and variability of tropospheric trace gases," *Tellus*, **26**, 477 (1974).

16. E. Robinson and R. C. Robbins, *Sources, Abundance and Fate of Gaseous Atmospheric Pollutants*, Standord Research Project PR-6755 (American Petroleum Institute, Washington, D.C., 1968).
17. L. A. Ripperton, H. E. Jeffries, J. R. Smith and C. E. Decker, private communication, as reported by Seiler.[5]
18. J. W. Swinnerton and R. A. Lamontagne, "Carbon monoxide in the South Pacific Ocean," *Tellus*, **26**, 136 (1974).
19. R. E. Newell, "One-dimensional models: a critical comment, and their application to carbon monoxide," *J. Geophys. Res.*, **82**, 1449 (1977).
20. W. L. Chameides, S. C. Liu and R. J. Cicerone, "Possible variations in atmospheric methane, *J. Geophys. Res.*, **82**, 1795 (1977).
21. J. W. Swinnerton, V. J. Linnenbom and C. H. Cheek, "A sensitive gas chromatographic method for determining carbon monoxide in sea water," *Limnol. Oceanogr.*, **13**, 193 (1968).
22. R. A. Lamontagne, J. W. Swinnerton, V. J. Linnenbom and W. Smith, "Methane concentrations in various marine environments," *J. Geophys. Res.*, **78**, 5317 (1973).
23. J. M. Brooks and W. M. Sackett, "Sources, sinks, and concentrations of light hydrocarbons in the Gulf of Mexico," *J. Geophys. Res.*, **78**, 5248 (1973).
24. R. A. Lamontagne, J. W. Swinnerton and V. J. Linnenbom, "Nonequilibrium of carbon monoxide and methane at the air–sea interface," *J. Geophys. Res.*, **76**, 5117 (1971).
25. V. J. Linnenbom, J. W. Swinnerton and R. A. Lamontagne, "The ocean as a source for atmospheric carbon monoxide," *J. Geophys. Res.*, **78**, 5333 (1973).
26. D. F. Wilson, J. W. Swinnerton and R. A. Lamontagne, "Production of carbon monoxide and gaseous hydrocarbons: relation to dissolved organic carbon," *Science*, **168**, 1577 (1970).
27. L. Winkler, "Gesetzmässigkeit bei der Absorption der Gase in Flüssigkeiten," *Z. Physik. Chem.*, **55**, 344 (1906).
28. E. Douglas, "Carbon monoxide solubilities in seawater," *J. Phys. Chem.*, **71**, 1931 (1967).
29. R. W. Meadows and D. J. Spedding, "The solubility of very low concentrations of carbon monoxide in aqueous solution," *Tellus*, **26**, 143 (1974).
30. W. J. Brtko and R. L. Kabel, "Pollutant transfer into water bodies," *Water, Air, Soil Poll.*, **6**, 71 (1976).
31. W. S. Broecker, Y. H. Li and T. H. Peng, "Carbon dioxide–man's unseen artifact," in *Impingement of Man on the Oceans*, edited by D. W. Hood (Wiley-Interscience, New York, 1971) pp. 287–324.
32. W. S. Broecker and T. H. Peng, "Gas exchange rates between air and sea," *Tellus*, **26**, 21 (1974).
33. A. Okubo, "Horizontal and vertical mixing in the sea, in *Impingement of Man on the Oceans*, edited by D. W. Hood (Wiley-Interscience, New York, 1971) p. 141.
34. C. Junge, W. Seiler and P. Warneck, "The atmospheric ^{12}CO and ^{14}CO budget," *J. Geophys. Res.*, **76**, 2866 (1971).
35. J. W. Swinnerton, V. J. Linnenbom and R. A. Lamontagne, "The ocean: a natural source of carbon monoxide," *Science*, **167**, 984 (1970).
36. T. Koyama, "Gaseous metabolism in lake sediments and paddy soils and the production of atmospheric methane and hydrogen," *J. Geophys. Res.*, **68**, 3971 (1963).
37. T. Koyama, "Biogeochemical studies on lake sediments and paddy soils and the production of atmospheric methane and hydrogen," in *Recent Researches in the Field of Hydrosphere, Atmosphere and Nuclear Geochemistry*, edited by Y. Miyake and T. Koyama (1964) pp. 143–177.
38. A. Baker-Blocker, T. M. Donahue and K. H. Mancy, "Methane Flux from wetland areas," *Tellus*, **29**, 245 (1977).
39. G. E. Hutchinson, "Circular causal systems in ecology," *Ann. N.Y. Acad. Sci.*, **50**, 221 (1948).
40. G. M. Woodwell, "The energy cycle of the biosphere," *Sci., Am.*, **223**, 64 (1970).
41. D. R. Hitchcock and A. E. Wechsler, *Biological Cycling of Atmospheric Trace Gases: Final Report*, NASW-2128 (1972).

42. R. E. Inman and R. B. Ingersoll, "Uptake of carbon monoxide by soil fungi," *J. Air Poll. Cont. Assoc.*, **21**, 646 (1971).
43. R. E. Inman, R. B. Ingersoll and E. A. Levy, "Soil: a natural sink for carbon monoxide," *Science*, **172**, 1229 (1971).
44. R. B. Ingersoll and R. E. Inman, "Soil as a sink for atmospheric carbon monoxide," in *Sources, Sinks and Concentrations of Carbon Monoxide and Methane in the Earth's Environment*, Joint Meeting of the American Geophysical Union and the American Meteorological Society (1972).
45. R. B. Ingersoll, R. E. Inman and W. R. Fisher, "Soil's potential as a sink for atmospheric carbon monoxide," *Tellus*, **26**, 151 (1974).
46. K. H. Liebl, *The Soil as a Sink and Source for Atmospheric CO*, Masters' Degree Thesis (Max-Planck-Institut and University of Mainz, 1971); for a summary of this work, see Seiler.[5]
47. U.S. Environmental Protection Agency, *Monitoring and Air Quality Trends Report* (Office of Air and Waste Management, Research Triangle Park, North Carolina, 1976).
48. R. E. Newell, G. J. Boer and J. W. Kidson, "An estimate of the interhemispheric transfer of carbon monoxide from tropical general circulation data," *Tellus*, **26**, 101 (1974).
49. J. Pressman and P. Warneck, "The stratosphere as a chemical sink for carbon monoxide," *J. Atmos. Sci.*, **27**, 155 (1970).
50. S. C. Wofsy, J. C. McConnell and M. B. McElroy, "Atmospheric CH_4, CO and CO_2," *J. Geophys. Res.*, **77**, 4477 (1972).
51. J. C. McConnell, M. B. McElroy and S. C. Wofsy, "Natural sources of atmospheric CO," *Nature*, **223**, 187 (1971).
52. L. Machta, "Sources and sinks from profile data," in *Sources, Sinks and Concentrations of Carbon Monoxide and Methane in the Earth's Environment*, Joint Meeting of the American Geophysical Union and the American Meteorological Society (1972).
53. P. J. Crutzen, "Gas-phase nitrogen and methane chemistry in the atmosphere," in *Physics and Chemistry of the Upper Atmosphere*, edited by B. M. McCormac (Reidel, Dordrecht, Netherlands, 1973).
54. S. C. Liu, T. M. Donahue, R. J. Cicerone and W. L. Chameides, "Effect of water vapor on the destruction of ozone in the stratosphere perturbed by ClX or NO_x Pollutants," *J. Geophys. Res.*, **81**, 3111 (1976).
55. R. S. Stolarski and R. J. Cicerone, "Stratospheric chlorine: a possible sink for ozone," *Can J. Chem.*, **52**, 1610 (1973).
56. D. Butler, private communication (1978).
57. S. Liu, private communication (1978).
58. B. Weinstock, "Carbon monoxide: residence time in the atmosphere," *Science*, **166**, 244 (1969).
59. H. Levy, "Normal atmosphere: large radical and formaldehyde concentrations predicted," *Science*, **173**, 141 (1971).
60. P. J. Crutzen, "Photochemical reactions initiated by and influencing ozone in unpolluted tropospheric air," *Tellus*, **26**, 47 (1974).
61. W. L. Chameides and J. C. G. Walker, "A photochemical theory of tropospheric ozone," *J. Geophys. Res.*, **78**, 8751 (1973).
62. W. L. Chameides and D. H. Stedman, "Tropospheric ozone: coupling transport and photochemistry," *J. Geophys. Res.*, **82**, 1787 (1977).
63. R. W. Stewart, S. Hameed and J. P. Pinto, "Photochemistry of tropospheric ozone," *J. Geophys. Res.*, **82**, 3134 (1977).
64. J. Fishman and P. J. Crutzen, "A numerical study of tropospheric photochemistry using a one-dimensional model," *J. Geophys. Res.*, **82**, 5897 (1977).
65. R. D. Hudson, (editor), *Chlorofluoromethanes and the Stratosphere*, NASA Reference Publication 1010 (Goddard Space Flight Center, Greenbelt, Maryland, 1977).
66. J. Heicklen, "Gas-phase reactions of alkyl peroxy and alkoxy radicals," *Adv. Chem. Ser.*, **76**, 23 (1968).
67. H. Levy, "Photochemistry of minor constituents in the troposphere," *Planet. Space Sci.*, **21**, 575 (1973).

68. C. T. Pate, B. J. Finlayson and J. T. Pitts, Jr., "A long path infrared spectroscopic study of the reaction of methylperoxy free radicals with nitric oxide," *J. Am. Chem. Soc.*, **96**, 6554 (1974).
69. K. L. Demerjian, J. A. Kerr and J. G. Calvert, "The mechanism of photochemical smog formation," *Adv. Env. Sci. Tech.*, **4**, 1 (1974).
70. J. G. Calvert, J. A. Kerr, K. L. Demerjian and R. D. McQuigg, "Photolysis of formaldehyde as a hydrogen atom source in the lower atmosphere," *Science*, **175**, 751 (1972).
71. G. R. McMillan and J. G. Calvert, "Gas phase photooxidation," *Oxid. Comb. Rev.*, **1**, 83 (1965).
72. W. H. Chan, W. M. Uselman, J. G. Calvert and J. H. Shaw, "The pressure dependence of the rate constant for the reaction: $HO + CO \rightarrow H + CO_2$," *Chem. Phys. Lett.*, **45**, 240 (1977).
73. D. D. Davis, W. A. Payne and L. J. Stief, "The hydroperoxyl radical in atmospheric chemical dynamics: reaction with carbon monoxide," *Science*, **179**, 280 (1973).
74. N. R. Greiner, "Hydroxyl radical kinetic spectroscopy. V. Reactions with hydrogen and carbon monoxide in the range 300–500 K," *J. Chem. Phys.*, **51**, 5049 (1969).
75. J. A. Davidson, C. M. Sadowski, H. I. Schiff, G. E. Streit, C. J. Howard, D. A. Jennings and A. L. Schmeltekopf, "Absolute rate constant determinations for the deactivation of $O(^1D)$ by time resolved decay of $O(^1D) \rightarrow O(^3P)$ emission," *J. Chem. Phys.*, **64**, 57 (1976).
76. D. L. Baulch, D. D. Drysdale, D. G. Horne and A. C. Lloyd, *Evaluated Kinetic Data for High Temperature Reactions*, Vol. 2, *Homogeneous Gas Phase Reactions of the* H_2–N_2–O_2 *System* (Chemical Rubber Co. Press, Cleveland, Ohio, 1973).
77. R. F. Hampson, D. Garvin, J. T. Herron, R. E. Huie, M. J. Kurylo, A. H. Laufer, H. Okabe, M. D. Sheer and W. Tsang, *Chemical Kinetics Survey. VI. Photochemical and Rate Data for Twelve Gas Phase Reactions of Interest for Atmospheric Chemistry*, NBSIR 73-207 (National Bureau of Standards, Washington, D.C., 1973).
78. L. T. Molina, S. D. Schinke and M. J. Molina, "Ultraviolet absorption spectrum of hydrogen peroxide vapor, *Geophys. Res. Lett.*, **4**, 580 (1977).
79. C. J. Howard and K. M. Evenson, "Kinetics of the reaction of HO_2 with NO," *Geophys. Res. Lett.*, **4**, 437 (1977).
80. C. J. Howard, "Kinetics of the reaction of HO_2 with NO_2," submitted to *J. Chem. Phys.* (1978).
81. R. A. Graham, A. M. Winer and J. N. Pitts, Jr., "Temperature dependence of the unimolecular decomposition of pernitric acid and its atmospheric implications," *Chem. Phys. Lett.*, **51**, 215 (1977).
82. H. Johnston and R. Graham, "Gas-phase ultraviolet absorption spectrum of nitric acid vapor," *J. Phys. Chem.*, **77**, 62 (1973).
83. D. Garvin and R. F. Hampson (editors), *Chemical Kinetics Data Survey. VII. Tables of Rate and Photochemical Data for Modeling of the Stratosphere*, NBSIR 74-430 (National Bureau of Standards, Washington, D.C., 1974).
84. A. B. Harker and H. S. Johnston, "Photolysis of nitrogen dioxide to produce transient O, NO_3 and N_2O_5," *J. Phys. Chem.*, **77**, 1153 (1973).
85. W. H. Chan, R. J. Nordstrum, J. G. Calvert and J. H. Shaw, "An IRFTS spectroscopic study of the kinetics and the mechanism of the reactions in the gaseous system HONO, NO, NO_2, H_2O," *Chem. Phys. Lett.*, **37**, 441 (1976).
86. E. D. Morris, Jr. and H. Niki, "The reaction of N_2O_5 with H_2O," *J. Phys. Chem.*, **77**, 1929 (1973).
87. M. Griggs, "Absorption coefficients of ozone in the ultraviolet and visible regions," *J. Chem. Phys.*, **49**, 857 (1968).
88. S. Kuis, R. Simonaitis and J. Heicklen, "Temperature dependence of the photolysis of ozone at 3130 Å," *J. Geophys. Res.*, **80**, 1328 (1975).
89. R. E. Huie, J. T. Herron and D. D. Davis, "Absolute rate constants for the reaction $O + O_2 + M \rightarrow O_3 + M$ over the temperature range 200–346°K," *J. Phys. Chem.*, **76**, 2653 (1972).
90. W. Jacobi and K. Andre, "The vertical distribution of radon 222, radon 220, and their decay products in the atmosphere," *J. Geophys. Res.*, **68**, 3799 (1963).

91. L. Machta, R. J. List, M. E. Smith, Jr. and H. Oeschger, "Use of natural radioactivities to estimate large-scale precipitation scavenging," *Precipitation Scavenging, A.E.C. Symp. Ser.*, **22**, 465 (1970).
92. P. Fabian, "A theoretical investigation of tropospheric ozone and stratospheric–tropospheric exchange processes," *Pure Appl. Geophys.*, **106–108**, 1044 (1973).
93. *U.S. Standard Atmosphere*, prepared under the sponsorship of NOAA, NASA and USAF (Washington, D.C., 1976).
94. *U.S. Standard Atmosphere Supplement*, prepared under the sponsorship of ESSA, NASA and USAF (Washington, D.C., 1966).
95. C. C. Wang, L. I. Davis, Jr., C. H. Wu, S. Japar, H. Niki and B. Weinstock, "Hydroxyl radical concentrations measured in ambient air," *Science*, **189**, 797 (1975).
96. D. D. Davis, W. Heaps and T. McGee, "Direct measurement of natural tropospheric levels of OH *via* an airborne tunable dye laser," *Geophys. Res. Lett.*, **3**, 331 (1976).
97. D. Perner, D. H. Ehhalt, H. W. Patz, U. Platt, E. P. Roth and A. Volz, "OH-Radicals in the lower troposphere," *Geophys. Res. Lett.*, **3**, 466 (1976).
98. W. L. Chameides, "The photochemical role of tropospheric nitrogen oxides," *Geophys. Res. Lett.*, **5**, 17 (1978).
99. J. F. Noxon, "NO_2 in the stratosphere and troposphere by ground based absorption spectroscopy," *Science*, **189**, 547 (1975).
100. H. B. Singh, "Atmospheric halocarbons: evidence in favor of reduced hydroxyl radical concentration in the troposphere," *Geophys. Res. Lett.*, **4**, 101 (1977).
101. B. K. T. Sie, R. Simonaitis and J. Heicklen, "The reaction of OH with CO," *Int. J. Chem. Kinetics*, **8**, 85 (1976).
102. R. A. Cox, R. G. Derwent and P. M. Holt, "Relative rate constants for the reactions of OH radicals with H_2, CH_4, CO, NO and HONO at atmospheric pressure and 296 K," *J. Chem. Soc. Faraday Trans.*, I, **9**, 2031 (1976).
103. C. E. Junge, *Air Chemistry and Radioactivity* (Academic, New York, 1963).
104. B. Weinstock and T. Y. Chang, "The steady state concentration of carbon monoxide in the troposphere," in *Proceedings of the Second International Conference on Environmental Biogeochemistry* (1975).
105. U.S. Public Health Service, *Air Quality Criteria for Hydrocarbons* (Department of Health, Education and Welfare, National Air Pollution Control Administration, Washington, D.C., 1970).
106. L. A. Cavanaugh, C. F. Schadt and E. Robinson, "Atmospheric hydrocarbon and carbon monoxide measurements at Point Barrow, Alaska," *Env. Sci. Technol.*, **3**, 251 (1969).
107. Research Triangle Institute, *Investigation of Ozone and Ozone Precursor Concentrations at Nonurban Locations in the Eastern United States*, EPA-450/3-74-034 (U.S. Environmental Protection Agency, Washington, D.C., 1974).
108. R. A. Rasmussen and E. Robinson, *Atmospheric Hydrocarbon Concentrations in Global Background Air Samples*, 77-13-48 (College of Engineering, Washington State University, Pullman, Washington, 1977).
109. R. N. Grenda and H. W. Goldstein, "Nonurban measurements of methane and ethane under varying atmospheric conditions," *J. Geophys. Res.*, **82**, 5923 (1977).
110. W. L. Chameides and R. J. Cicerone, "Effects of nonmethane hydrocarbons in the atmosphere," *J. Geophys. Res.*, **83**, 947 (1978).
111. G. Czeplak and C. Junge, "Studies of interhemispheric exhcange in the troposphere by a diffusion model, *Adv. Geophys.*, **188**, 57 (1974).
112. H. C. W. Kwok, W. E. Langlois and R. A. Ellefsen, "Digital simulation of the global transport of carbon monoxide," *I.B.M. J. Res. Develop.*, **15**, 3 (1971).
113. L. K. Peters and A. A. Jouvanis, "Numerical simulation of the transport and chemistry of CH_4 and CO in the troposphere" *Atmos. Environ.* (1979) in press.
114. R. H. Kummler and T. Baurer, "A temporal model of tropospheric carbon–hydrogen chemistry," *J. Geophys. Res.*, **78**, 5306 (1973).
115. B. Weinstock and H. Niki, "Carbon monoxide balance in nature," *Science*, **176**, 290 (1972).
116. Committee on Interstate and Foreign Commerce, House of Representatives, *Fluoro-*

carbons—Impact on Health and Environment, Hearings before Subcommittee on Public Health and Environment, 93-110 (U.S. Government Printing Office, Washington, D.C., 1974).

117. W. C. Wang, Y. L. Yung, A. A. Lacis, T. Mo and J. E. Hansen, "Greenhouse effects due to anthropogenic perturbations," *Science*, **194**, 685 (1976).
118. N. D. Sze, "Anthropogenic CO emissions: implications for the atmospheric CO–OH–CH_4 cycle, *Science*, **195**, 673 (1977).
119. W. L. Chameides, "The effect of anthropogenic carbon monoxide on the methane budget of the troposphere," in *Proceedings of the Fourth Joint Conference on Remote Sensing of Environmental Pollutants* (American Chemical Society, 1978).
120. S. C. Liu, "Possible effects on the tropospheric O_3 and OH due to NO emissions," *Geophys. Res. Lett.*, **4**, 325 (1977).
121. U.S. Public Health Service, *Air Quality Criteria for Carbon Monoxide* (Department of Health, Education and Welfare, National Air Pollution Control Administration, Washington, D.C., 1970).
122. U.S. Environmental Protection Agency, *Air Quality Criteria for Nitrogen Oxides* (Air Pollution Control Office, Washington, D.C., 1971).
123. L. Margulis and J. E. Lovelock, "Biological modulation of the Earth's atmosphere," *Icarus*, **21**, 471 (1974).

Leonard K. Peters
University of Kentucky
and William L. Chameides
University of Florida

RURAL OZONE—SOURCES AND TRANSPORT

In the years following the discovery of ozone by Schönbein in 1840,[1] a number of workers set out to measure the ozone content of the atmosphere. They used the "Schönbein method"; that is, they dipped pieces of filter paper in a solution of starch and potassium iodide (KI) and hung these in the open air. As the ozone released iodine (I_2) from the iodide, the iodine entered the helices of the starch molecules, thus producing a blue color.

Schönbein's method was relative and was based on seven intensities of blue. The early workers soon found that the ozone was apparently more concentrated in country air than in city air. Many workers decided that ozone was a healthful substance, presumably because the countryside was inhabited by luscious lassies and lusty laddies, whereas the city was inhabited by pale consumptive creatures lurking in the slums and around the chemical plants. Ozone became the name of health spas and other places hoping to attract visitors and was incorporated into the names of several communities in Europe and in the United States, as witness the names of Ozone Park, New York; Ozona, Florida; Ozona, Texas; and Ozone, Tennessee.

What was not realized at first was that sulfur dioxide would interfere with Schönbein's analytical technique on a one-to-one basis. Sulfur dioxide was a virtually ubiquitous substance in the air of any coal-burning city, and so we cannot place much reliance on these early measurements and conclusions.

Ozone (O_3) in the troposphere can be said to have three general origins. By far the greatest quantity of ozone associated with the Earth's atmosphere is generated in the stratosphere by a series of reactions initiated by the photolysis of molecular oxygen (O_2). This is then followed by transfer across the tropopause into the troposphere. This contributes to the so-called natural background of O_3 in the troposphere.

A set of chemical reactions initiated by the photolysis of nitrogen dioxide (NO_2), the NO generated being subsequently oxidized back to NO_2 by some species other than O_3, provides the basis of tropospheric synthesis of O_3. Undoubtedly, small concentrations of O_3 are generated in the troposphere from naturally occurring NO_2 and organic compounds, and form part of the natural background concentration of O_3. Finally, the same generalized set of reactions occurs in urban air, where the precursor concentrations are much higher. Much higher concentrations of O_3 can be generated than are found in natural background air.

There is one event which, though it has not been studied, can be responsible for very high concentrations of "naturally" occurring O_3. Forest fires generate great amounts of NO_x and distil large quantities of organic compounds from the wood. A study of an experimental burning of eucalyptus wood in Australia showed the generation of concentrations above the NAAQS in the downward plume.[2] This has brought up the question of what are natural concentrations. Interest has centered around the concentration 0.080 ppm; the reason is that in 1971 the National Ambient Air Quality Standard (NAAQS) was promulgated for photochemical oxidant of 0.080 ppm hourly average, not to be exceeded more than once a year.

It was at one time fairly well agreed that the natural background of O_3 hardly, if ever, rose above 0.060 ppm. However, some time between the early 1960s and the early 1970s, measurements equal to and greater than the NAAQS began to be reported from what had been considered remote sites. A number of people considered these relatively high concentrations to be due to stratospheric intrusion. Others thought that they were due to intrusion of urban air or to what was called a spent pollution system from urban areas. Our belief, backed by examination of the data, is that indeed those areas that have been studied are experiencing a degradation of air quality as a result of the transport of anthropogenic pollution. We have concluded that the apparent discrepancy is due to a relatively sharp rise in surface O_3 concentration as a result of anthropogenic pollution, which occurred between the early 1960s and the early 1970s.

The consensus about the natural concentrations and behavior of O_3 derived from the earlier measurements was drawn from Junge's review of surface O_3 measurements from data taken at Arosa, Switzerland[3]; measurements made by Regener in Nebraska and New Mexico[4]; and more recent measurements by a number of workers,[5–10] as well as those made by the Research Triangle Institute/University of North Carolina team, which included 5 years of surface measurements in Piedmont North Carolina; observations made in the Appalachians of North Carolina and West Virginia; observations from Pikes Peak, Colorado; and observations from the Atlantic Ocean off the coast of North Carolina, and from the Gulf of Mexico and Caribbean areas.

Figure 1 is fairly representative of this natural background in the midlatitudes, in valleys or on level ground. The range of values of O_3 concentrations is fairly well represented by Table I, the summary made by Junge in 1963.[3]

TABLE I

Concentrations of tropospheric ozone as given in more recent and extensive studies (from Junge[3])

Source	Location, time and remarks	Altitude	O_3 (ppm) range	average
Götz and Volz (1951)	Arosa, Switzerland, 1950–51, high valley, daily maxima values	1860 m	0.010–0.045	0.025
Regener[4] (1957)	Mt. Capillo and Albuquerque, New Mexico, 1951–52	3100 m	0.009–0.043	0.023
		1600 m	0.002–0.061	0.018
Regener[4] (1957)	O'Neil, Nebraska, 1953	12.5 m above ground	0.015–0.051	0.031
Ehmert (1952)	Weissenau, Bodensee, Germany, 1952	20 m above ground	0.000–0.046	0.018
			0.000–0.036	0.015
Teichert (1955)	Lindenberg Observatory, Germany, 1953–54	80 m above ground	0.000–0.026	0.015
			0.000–0.026	0.014
Kay (1953)	Farnborough, England, 1952–53	0–12,000 m	0.013–0.026	0.019
Brewer (1955)	Tromsö, Norway, 1954	0–10,000 m	0.031–0.036	0.033
Price and Pales (1959)	Mauna Loa Observatory, Hawaii	3000 m	0.015–0.032	0.023
Wexler *et al.* (1960)	Little America Station, Antarctica	100 m	0.010–0.031	0.023

[a] As interpreted from the published data. The values sometimes represent absolute maxima, sometimes mean maxima.

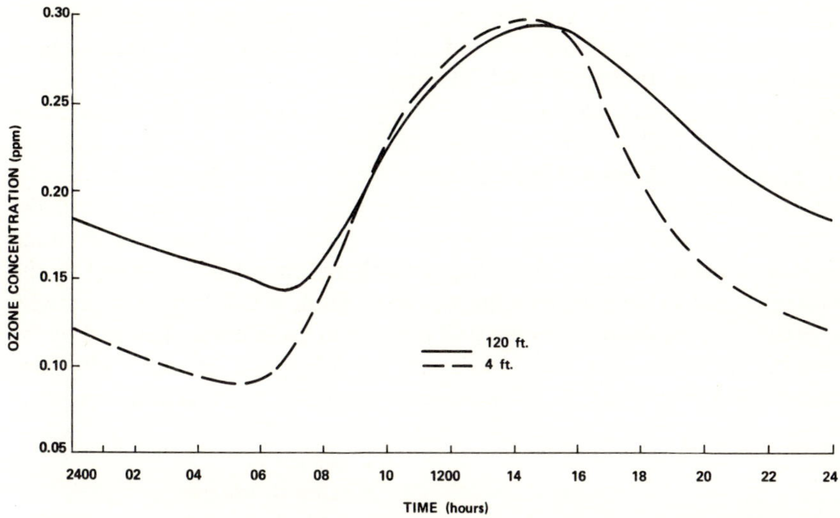

Figure 1 Average tower diurnal ozone concentrations (Piedmont, North Carolina).

In the studies from flatlands and valleys, the maximum concentration generally occurred shortly after noon, and decreased during the late afternoon to a minimum at night or just at dawn. One difference between rural air at remote sites and urban air was that the urban air generally had higher maximum concentrations and more frequently dropped to a measured zero at night, presumably as a result of the emission of O_3-destructive agents such as NO and NO_2 and certain organic compounds. In most of the observed measurements, there appeared to be a radiation inversion established almost every night. This permits the establishment of two O_3 regimes, one at the ground (lower concentration) and one above the radiation inversion. Near the ground the ozone can be destroyed by contact with the ground and by O_3-destructive gases emanating from the surface. In the morning the sun hits the Earth and starts to dissipate the radiation inversion, thus mixing the surface air with air further aloft. This increases the concentration of O_3 at the surface by mixing the air from aloft with the surface air. In a study of the synthesis of O_3 in the natural surface air, it was possible to demonstrate the generation of a small concentration of O_3,[11] the apparent role of this synthesis being to help maintain the O_3 concentration and cut down on the net rate of O_3 destruction.

From a tower site in Piedmont, North Carolina (Figure 1), it was shown that on many mornings the O_3 concentration, which was generally higher at 120 ft than at 4 ft, became very slightly higher at 4 ft for a short time in the early morning hours, before the base of the dissipating radiation inversion had lifted above the 120-ft sampling inlet.[12] The explanation is that the O_3 precursors were more concentrated at the ground than at 120 ft. A brief experiment was conducted in which ambient air was run through MNO_2 to destroy the O_3 and then run into a 72-liter flask exposed to the natural sunlight. The air had a 30-min average residence time in the flask and was sampled as it left the flask. Figure 2 depicts the most striking results. Generally, the O_3 potential of the air as measured thus tended to decrease after 0900. Other evidence of the natural tropospheric synthesis of ozone was obtained by McKee in Greenland.[13] Sampling with phenolphthalein, he found that the oxidant in the air flowing from the ice cap at Thule was 0.06 ppm in sunlight, whereas it would drop to around 0.02 ppm when the sun was sheltered by the clouds. These findings were discussed by Ripperton *et al.*,[11] who also summarized other evidence of small-scale synthesis of ozone in natural air.

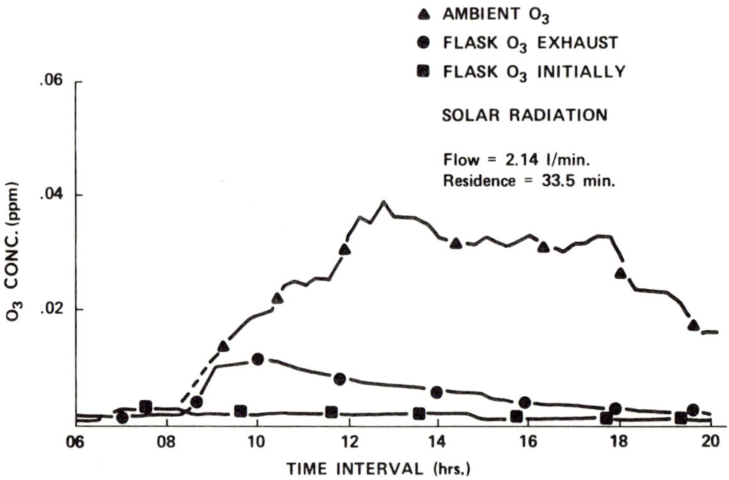

Figure 2 Ozone potential of ambient atmosphere.

Surface ozone in the Northern Hemisphere, at least in the midlatitudes, seems to peak in concentration in April. In contrast to the ozone measurement over land, ship-borne measurements taken at sea did not produce a diurnal variation typical of flat land. The maximum observed on the Atlantic was approximately 0.05 ppm, while in the more southerly waters the maximum was much less.[14] The lack of diurnal variation at sea can be explained by the fact that a radiation inversion is seldom if ever formed in those waters. Thus, at night the depth of the mixing layer remains approximately the same. On mountain tops such as Green Knob (5160 ft)[15] and Pikes Peak (14,100 ft),[16] the diurnal trend observed on flatlands is reversed, and the peak of ozone concentration occurs at midnight or just at daybreak, with minimum concentrations at noon (Figure 3).

The graph containing the Green Knob data shows also the ozone-concentration data from the adjacent valley, 1600 ft below Green Knob. The explanation for the inverse relationship between peak and valley is that, in the middle of the day, the mixing layer includes air from both sampling sites, the air with the higher concentrations from aloft and the air originally containing lower concentrations from below. Thus, the concentrations are approximately equal. When the radiation inversion sets in, the air on the peak becomes cooler and denser and is subject to a downslope flow. This flow travels down the mountain, mixing with gaseous emanations from the vegetation, and results in ozone destruction and a subsequent reduction in the surface ozone concentration. The air flowing downward from the peak is replaced by air from the free atmosphere, which contains the higher concentrations of O_3 measured at the peak. The O_3 measured in the valley is depleted because of contact with the mountainside and its vegetation.

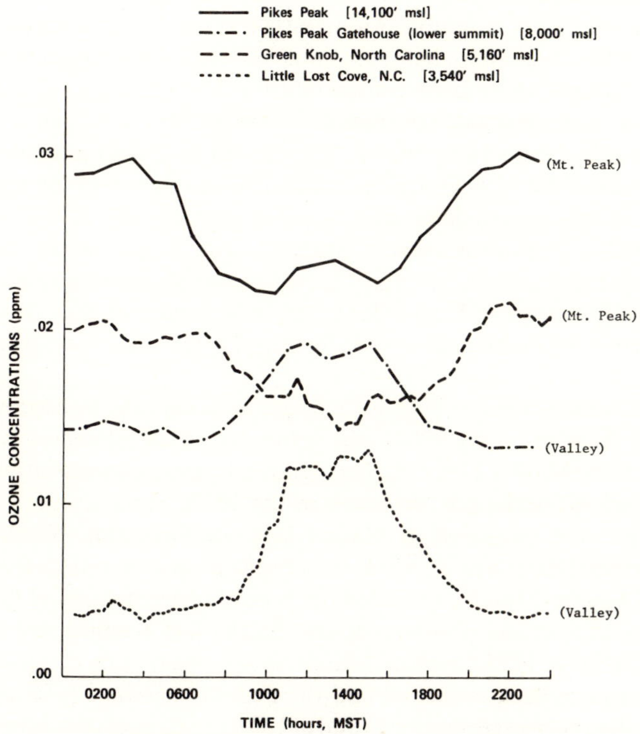

Figure 3 Comparison of mountain-peak and valley diurnal ozone data.

Presumably the higher nocturnal mountain top concentrations, which are almost constant with time, represent the background concentration of ozone in the air mass moving into the area, whereas the dips in daylight hours represent the local ozone concentrations in contact with vegetation.

In the early 1970s, but starting perhaps before then, reports began to trickle in of "high" concentrations of ozone in rural areas previously thought to be remote. This brought up the question of transport of pollutants, especially ozone and its precursors, over large distances. In 1958, Stanford Research Institute reported an increase in CO being measured in Greenland. After trajectory analysis, they ascribed this increase to air that had previously passed over industrial areas of the United States just south of the Great Lakes. In further support of the transport theory, a photograph taken by the crew of Skylab 4 (December 7, 1973) and published in the *National Geographic Magazine* shows a plume from a reed fire in Louisiana holding its visual integrity more than 200 miles into the Gulf of Mexico. The concept of long-distance transport of air pollution is quite widely accepted today.

In the late 1960s, damage to Christmas-tree plantations in northeastern West Virginia and western Maryland occurred, which was thought to have been caused by air pollution. An investigation and field-measurement program (for O_3, NO, NO_2, SO_2, etc.) was initiated. Although detailed investigations brought about a concensus that the damage was caused by SO_2, it was noted that the ozone concentration was inordinately high for what was considered to be a remote region. Subsequent field studies established that the sources of the tree-damaging pollutants and the observed high concentrations of ozone were different.

In 1970 a field-measurement program was conducted to confirm the presence of high concentrations of rural ozone.[17] Some of the data from this study are presented in Figure 4a. Later, field work was done to attempt to characterize the high-ozone system further. Typical data are shown in Figure 4b.

The literature was searched for references to high ozone concentration in rural air from the early 1960s and before; no such references were found. Junge's 1959 Mauna Loa data had reported ozone concentrations ranging from 0.015 to 0.031 ppm; however, in the 1970s concentrations as high as 0.09 ppm were measured at Mauna Loa. In Piedmont, North Carolina, rural-ozone data taken in 1966–67, covering an 8-month fall and winter period (August–March), exceeded the hourly concentration of 0.08 ppm 24 times ($\sim$0.5% of the observed hours). In the last 8-month fall and winter period (August 1977 through March 1978), maximum concentrations of 0.11 ppm have been measured and the Federal standard (0.08 ppm hourly average) has been exceeded for a total of 130 h ($\sim$2.2% of the observed hours). Both the 1966–67 and 1977–78 data were measured from the same area at the

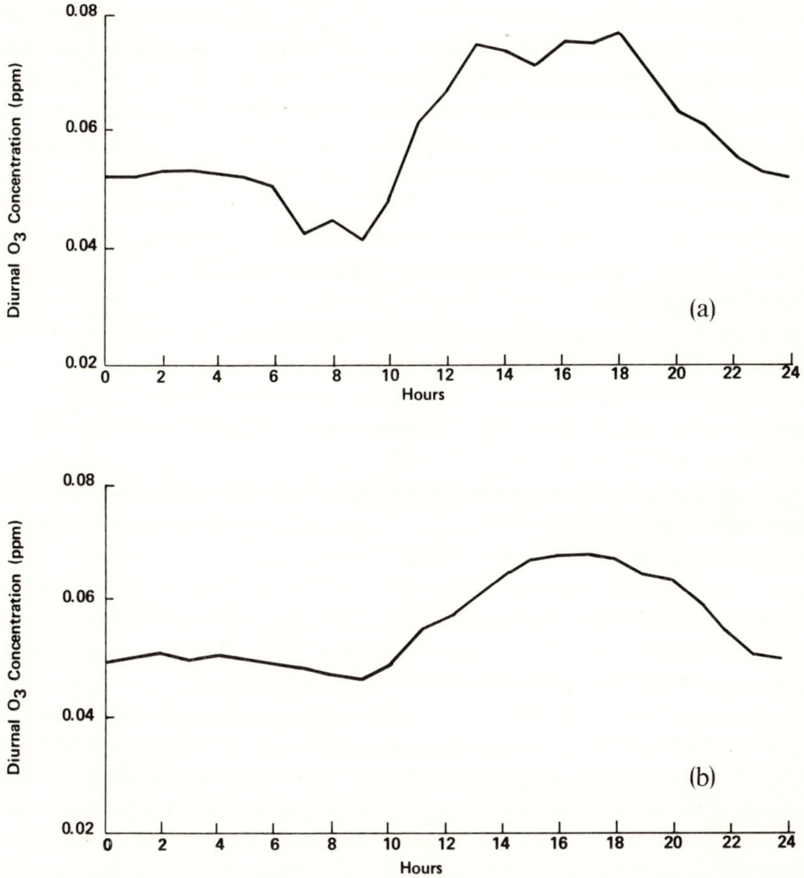

Figure 4 (a) Mean diurnal ozone variation in Mt Storm area (Steyer No. 2) between August 26 and September 9, 1970; (b) mean diurnal ozone concentration at Garrett County Maryland Airport between August 4 and September 25, 1972.

Research Triangle Institute rural facility. Viewed in this light, the ozone soundings reported by Hering and Borden[18] are quite interesting, in that the earlier soundings, when compared with later soundings, show increased encroachment of ozone-laden urban air into some of the surrounding regions in the later measurements.

We have concluded that some sort of quantitative change in ozone concentration has taken place in many parts of the United States. None of the meteorological, climatological or geophysical variables that are routinely measured correlate with the suggested increases in ozone concentration. As to the precursors of tropospherically generated ozone, the anthropogenic hydrocarbon emissions between 1960 and 1970 actually decreased, but the anthropogenic output of NO_x doubled.[19] This does not say that the increase in NO_x was responsible for the increased ozone, but it will bear investigation. In a recent report, Singh noted a number of remote sites, including Mauna Loa, reporting occasional concentrations equal to or greater than 0.08 ppm.[20] It is often dangerous to ascribe such concentrations to natural processes without an investigation to discover the source of the ozone, for example, by means of soundings or aircraft. The fact that some of these high concentrations occurred in February makes the possibility that they are natural seem greater.

A word or two is appropriate about the chemistry of the tropospheric synthesis of ozone. The simplest statement that hydrocarbons and NO_x plus sunlight equal smog does not give too accurate a picture of the behavior of ozone in the polluted air. NO_2 is photolyzed to yield $NO + O$; $O + O_2 + M \rightarrow O_3 + M$; $O_3 + NO \rightarrow NO_2 + O_2$.

This process is cyclic and eventually yields no net ozone. If, however, there are organic vapors present, the organic molecules whose oxidation is begun by the atomic oxygen will, in a highly complex mixture of reactions, yield organic peroxy-free radicals, which will oxidize the NO to NO_2, thereby allowing the accumulation of an ozone molecule. NO and NO_2 can react with and destroy ozone; however, the normal photochemical reactions are producing ozone at a faster rate than it is being destroyed.

In the dark, when there are no photochemical reactions, the destruction of ozone by NO and NO_2 continues. Thus, in the city at night when NO and NO_2 are continuously being emitted, the ozone concentration frequently drops to a measured zero. As a matter of fact, the O_3–NO reaction is about 500 times faster than most ozone–olefin reactions. The O_3–NO_2 reaction is about as fast as the ozone–olefin reaction. Therefore, it must be realized that most of the ozone precursors are also ozone-destructive. For practical purposes, it may be said that the saturated hydrocarbons are not ozone-destructive.

The NO_x and the olefinic hydrocarbons are greatly depleted in a photochemical urban pollution system as it moves into the surrounding countryside. The NO_x is frequently not present in concentrations high enough to destroy the ozone. The NO_2 is frequently present in the low-ppb range.

At a constant concentration of 3 ppb, NO_2 would be photolyzed rapidly enough to form a theoretical 0.084 ppm in 1 hr at midday. This would, of course, require that none of the ozone be destroyed.

The saturated hydrocarbons, which do not react readily with ozone, can nevertheless be oxidized in such a spent system to form peroxy-free radicals. The rate of formation of the peroxy-free radicals will determine how much of the theoretical ozone is actually realized.[21]

As a photochemical system is diluted by mixing with surrounding country air, it becomes more efficient in ozone synthesis. For example, if a system is diluted by 50%, the potential ozone concentration derived from the irradiation of the system may be decreased, but it will be decreased by less than 50%. The reduction of the concentration of ozone-destructive agents in a spent system also allows the transport of high concentrations of ozone for great distances aloft and at night.

Smog-chamber studies in which reaction mixtures were irradiated with natural sunlight for three daylight periods simulated the behavior of rural ozone concentrations in the second and third days, confirming the dilution effect and the generation of concentrations above the NAAQS from NO_x concentrations of a few ppb.[22] Further field confirmation of the so-called dilution effect can be seen from California data in which the sampling station at Indio, a town of about 14,000, records a greater frequency of ozone values equal to or over 0.10 ppm than does Los Angeles.[23] The photochemical-pollution systems reaching Indio most probably come from the Los Angeles basin.

A field study was made to characterize more fully the high-ozone system measured in 1970.[24] The study was confined to an area of some 500 square miles. The base sampling station was located at the Garrett County Airport, just north of McHenry, Maryland, and samples were also taken at eight satellite stations surrounding the centrally located airport site. The field program ran from August 4 through September 25, 1972, and produced what was considered at that time a very unusual spectrum of rural ozone measurements.

During the 52-day field program, hourly ozone concentrations exceeded the Federal standard of 0.08 ppm 11.6% of the time, reaching a single-hour maximum of 0.12 ppm. The longest continuous period of high ozone ran for 26 h, beginning before daybreak on September 2.

The random distribution of ozone, both high and low, correlated poorly with wind direction and seemed to suggest a broad background phenomenon on a scale larger than the sampling area. In order to identify broad background patterns of rural ozone more clearly, another study was conducted in 1973.[25] This study employed four rural ground stations at Coshocton, Ohio; Kane, Pennsylvania; Lewisburg, West Virginia; and the 1972 location at McHenry, Maryland; altogether, the areal coverage was around 30,000 square miles. For the first time, an air-borne sampling program was used to provide a two-dimensional picture of the distribution of ozone over the study area.

One such flight on September 12, 1973, departed Raleigh/Durham at 0955 and proceeded north at an altitude of 8500 ft msl passing over Lewisburg, McHenry and Kane before turning southwest to fly over Coshocton and returning to Raleigh/Durham at 1755 that evening. The highest concentrations of ozone measured on this flight occurred in the heavily industrialized area around Pittsburgh. In order to link the 8500-ft ozone distribution to ground measurements, the aircraft completed a descent–ascent maneuver and a fly-by at 50 ft at each of the ground stations. Adding this third dimension to the model produced some significant findings (Figure 5), notably a dramatic vertical separation of higher ozone values near the surface and lower values at the 8500-ft flight altitude. As it turned out, the aircraft was flying not only above the mixing layer but also above a subsidence inversion associated with a high-pressure system that dominated the study area on September 12. This flight strongly suggested that the vertical distribution of ozone in the lower troposphere is linked to normal meteorological processes in the mixing layer and can be significantly altered by a subsidence inversion at the top of the mixing layer.

Analysis of the 1973-study data suggested two other significant relationships; the rural–diurnal ozone-concentration pattern and the high-ozone–high-atmospheric-pressure link. The typical urban–diurnal ozone-concentration curve had previously been established, and the rural curve showed similarities in shape, with a predawn minimum rising to a midafternoon maximum. The only major difference seemed to be that, although the typical night-time minimum for the city was near zero, the typical rural–diurnal minimum held at some value significantly above zero. Further, there was some 50% difference in the average minimum rural values (0.03 ppm for Lewisburg and 0.06 ppm for McHenry), suggesting that the characteristics of the ambient atmosphere associated with each site were different.

Further analysis of the 1973 data indicated that periods of increased atmospheric pressure associated with the movement of high-pressure systems into the area were generally accompanied by an increase in the average ozone over the study area. This was the first hint that region-wide ozone concentrations might be linked to a synoptic-scale meteorological phenomenon.

A 1974 ozone study was similar in scale to the 1973 study and covered roughly the same geographical area.[26] The main thrust of sampling was in the northern Ohio Valley area, with manned rural stations at Wilmington, Wooster and McConnelsville, Ohio. For continuity with previous work, unmanned sites were maintained at Du Bois, Pennsylvania and McHenry, Maryland. Data were also obtained from six urban stations: Cincinnati, Dayton, Columbus, Cleveland, Canton and Pittsburgh.

The 1974 study was designed to focus on the variations in rural ozone concentrations and to examine specific influences, such as urban plumes, on the rural burden. It is of immediate interest that the five rural stations experienced more frequent, longer and slightly more severe ozone episodes than the six city stations. This can be partially explained by the night-time

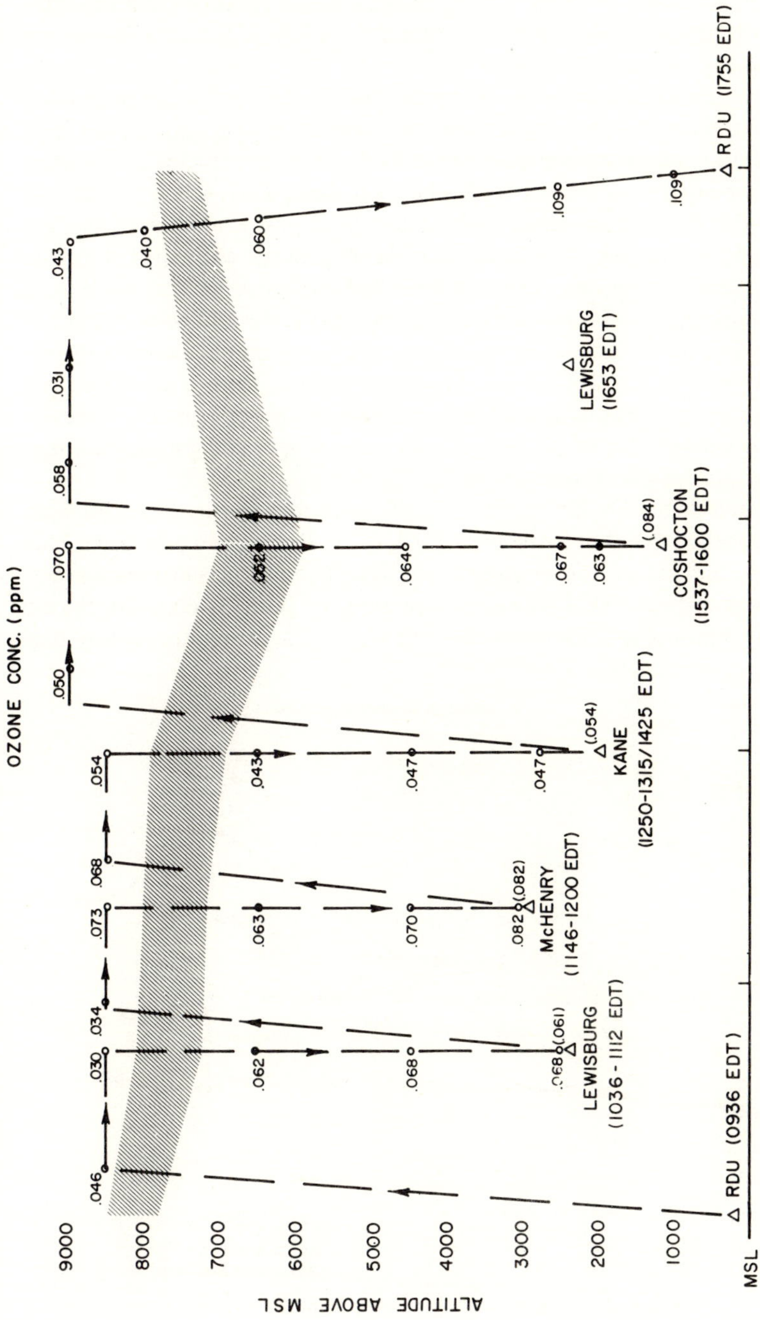

Figure 5 Aircraft measurements of ozone concentration (September 12, 1973).

destruction of ozone in cities, a phenomenon not typical of rural environments, so that more sustained ozone-burden buildups are possible in the rural areas.

In an attempt to isolate the mechanisms of ozone-precursor transport from urban to rural settings, two tools were employed extensively during the 1974 study—aircraft sampling and an air-mass-trajectory model. One of the more valuable air-borne operations took place on August 21, when an aircraft traversed the entire study area, making low-level passes at each of the five rural sampling stations to ensure that both ground and aircraft ozone monitors were compatible. The 700-mile flight (Figure 6) was made at a constant altitude of 4000 ft msl and revealed two scales of ozone-precursor transport. The more obvious were the several city plumes in evidence downwind of Wheeling and Pittsburgh. However, plumes from cities could not explain the concentrations approaching 0.2 ppm in the Du Bois area; here the air-mass-trajectory model produced evidence strongly suggesting the long-range transport of ozone precursors.

The air arriving at Du Bois at the time of the aircraft was in the region had, 48 h previously, passed over the New York–New Jersey industrial area. This area has the highest hydrocarbon emission density on the eastern seaboard—on the order of 500–1000 tons per square mile per year. By contrast, the air arriving at Wilmington, where ozone levels remained comparatively low, had

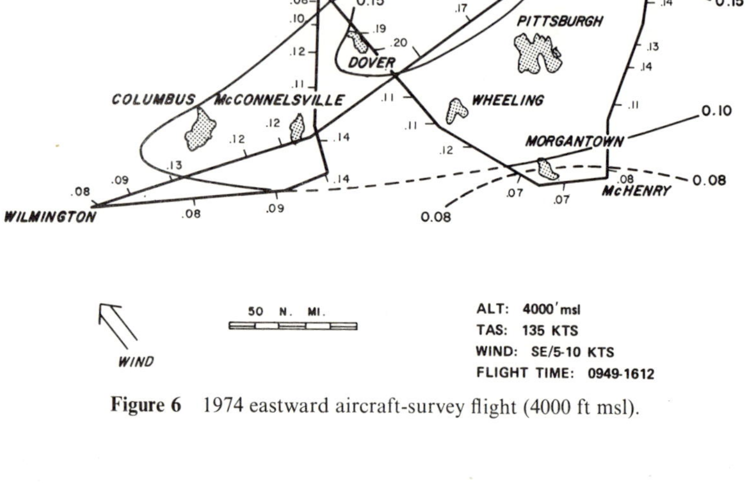

Figure 6 1974 eastward aircraft-survey flight (4000 ft msl).

traversed the relatively clean area of southern Virginia and central West Virginia. Generally speaking, the higher values of ozone measured on the August 21 flight occurred downwind of ozone precursor sources, with evidence pointing to both long- and short-range transport.

Another phase of the 1974 aircraft operations was devoted specifically to measuring the effects of cities on the rural area immediately around them. One such flight pattern (Figure 7) was a double-box configuration, centered on the urban area: the interior box was 25 nautical miles on a side, and the outer box 75 nautical miles on a side. Under conditions of high pressure, characterized by light winds and mild stagnation, the double-box pattern was flown around Columbus, Ohio, on July 9, revealing a small but discernible urban ozone plume superimposed on a relatively uniform ozone background

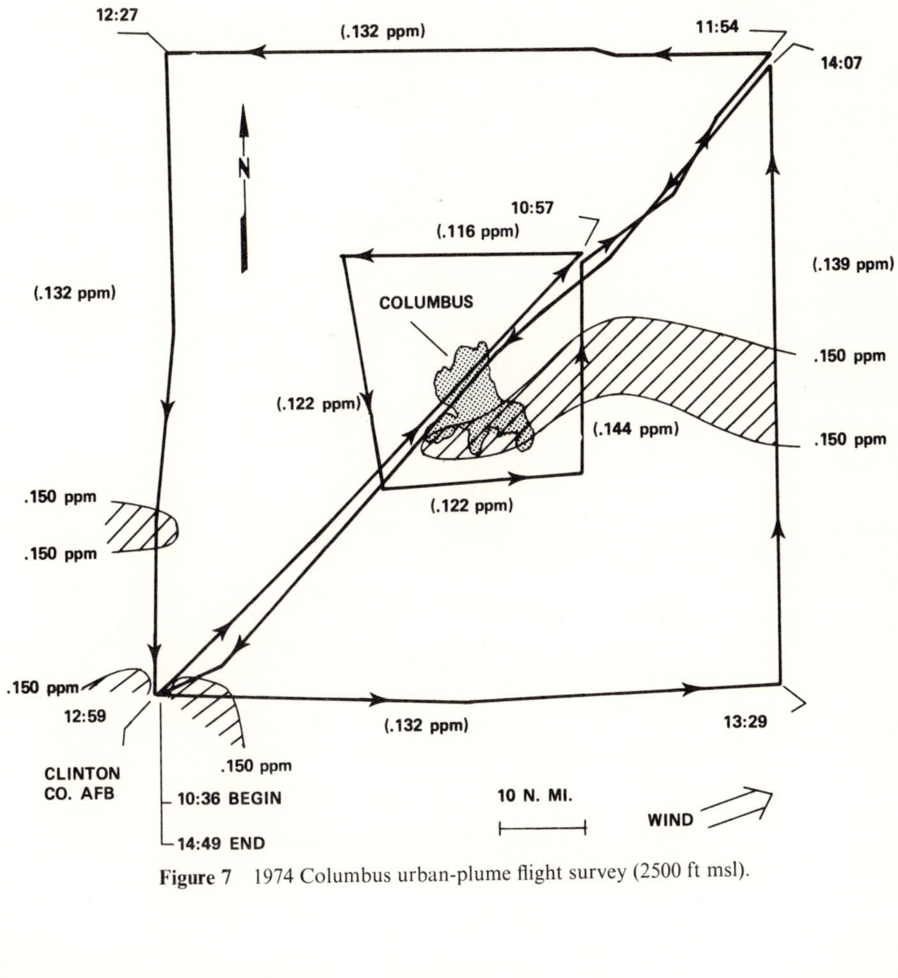

Figure 7 1974 Columbus urban-plume flight survey (2500 ft msl).

downwind to the east of the city. Two other ozone peaks near the southwestern corner of the outer box suggested the existence of similar plumes emanating from Dayton and Cincinnati.

In addition to the numerous aircraft missions devoted to measuring the horizontal ozone distribution, several vertical profiles were flown, the three most significant being executed in series on August 1 (Figure 8). Conditions that day were typical of those associated with a high-pressure system: strong nocturnal inversion at the surface, light winds throughout the mixing layer,

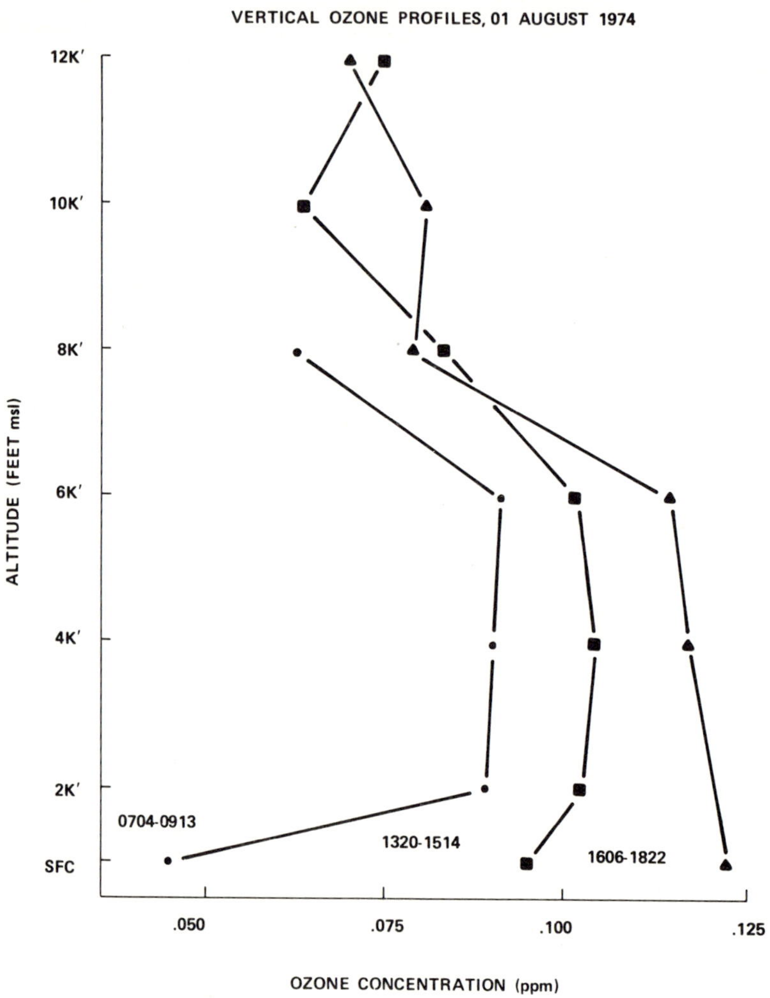

Figure 8 Vertical rural ozone profiles—Clinton County Airport, Ohio.

clear skies and a capping subsidence inversion aloft. The 0700 EDT profile graphically demonstrated the ozone-destructive effects of the 1000-ft-deep nocturnal surface inversion. The ozone concentrations in the layer between the top of the surface inversion at 2000 ft msl and the bottom of the subsidence inversion at 8000 ft msl were relatively constant at 0.09 ppm. Above the subsidence inversion the concentrations decreased sharply. The ozone concentration at the surface at 0700 was 0.05 ppm, the ozone had been destroyed during the night by contact with surface objects and vegetation, while the stable inversion layer prevented the downward mixing of the higher concentrations from the layer above. By midday (1300 EDT) the surface inversion had burned off, and ozone was mixed rather uniformly from the surface to the base of the subsidence inversion, with an average concentration of approximately 0.11 ppm. By later afternoon (1600 EDT), the mixing-layer concentration was a relatively uniform 0.12 ppm. With the subsidence inversion preventing a stratospheric intrusion of ozone and the light winds negating long-range transport, the August 1 flights presented a strong case for the daytime *in situ* buildup of ozone in the mixing layer in the presence of precursors that may have been transported into the area during the previous day.

In the next year, 1975, a study was conducted over an area ranging from Montana to the Atlantic seaboard.[27] This study area provided an area of steadily increasing population density with its associated pollution potential (an average of 1 person per square mile in the eastern Rockies, increasing to over 600 persons per square mile in the East Coast megalopolis). It is the area of the United States traversed by the major west–east moving weather system.

Three manned rural sampling sites—Wolf Point, Montana; Creston, Iowa; and Bradford, Pennsylvania—represented the west–east population gradient. The mean diurnal ozone curves for these three sites were similar to previous rural curves and differed systematically from one another only in the magnitude of the values: Bradford showed higher maximum, higher mean maximum, and higher mean minimum values (0.12 ppm, 0.06 ppm and 0.03 ppm, respectively) than Wolf Point (0.06 ppm, 0.04 ppm and 0.02 ppm), with Creston in between. This rural diurnal data strongly suggested higher ozone concentrations in areas with more anthropogenic ozone-precursor sources.

The mean diurnal ozone curves for the four city stations—Des Moines, Omaha, Columbus and Pittsburgh—were also typical, showing greater amplitude, higher maxima and lower minima than the rural curves. These measurements served to illustrate the destructive effects on ozone of the higher concentrations of ozone-scavenging pollutants in the urban atmosphere.

This study also allowed further documentation of the fluctuation of ozone burden with the passage of major weather systems. Using data from 1975 and the two previous years, a generalized model of this background ozone fluctuation was developed. Beginning with the passage of a cold front, the temporal variation of the diurnal ozone maximum roughly followed a sine curve. The model begins with a cold-front passage that triggers a 2-day decrease in ozone, followed by a 3-day increase to a peak as a high-pressure system moves through the area, the peak occurring about 2 days after the passage of the center of the anticyclone. The implication, borne out by the data, is that the longer the atmosphere is under the influence of high-pressure conditions, particularly radiation and limited vertical mixing, the higher will be the buildup of ozone.

The diurnal sequence of events at one of these sites under a high-pressure system seems to be that, as the high ozone system moves into an area in the afternoon, the ozone concentration is relatively uniform throughout the mixing layer. Then, as the nocturnal radiation inversion sets in, two ozone-concentration regimes are established: that in the radiation inversion and that above the radiation inversion. The ozone concentration above the radiation inversion drops very little, because it is the remainder of a relatively spent photochemical system with relatively low concentrations of NO_x and olefins. The ozone concentration in the radiation inversion is destroyed much more rapidly by contact with the ground and by reaction with surface emissions of NO_x and olefins. The next morning the surface concentration is somewhat lower than the concentration aloft but, as the solar heat burns off the radiation inversion, air from aloft containing more O_3 than that near the ground is mixed downward, raising the surface concentration. At the same time, photochemical reactions are initiated, and these generate additional ozone, thus producing essentially equal concentrations of ozone aloft and at the surface, rather higher than the value at either point at daybreak.

In 1976, Da Vinci, an urban-plume study program, was conducted in the St Louis area.[28] The primary sampling platform was the balloon-borne gondola (Da Vinci). It was planned that the balloon would be released into the St Louis urban plume and would drift in the plume for as long as possible. This Lagrangian sampling study was supported by aircraft flights that transected the flight tract of the balloon, and by instrumented mobile units that maintained continuous visual contact with the gondola from the ground and made simultaneous measurements. One of the study objectives of this experiment was to determine if a nocturnal inversion would separate the urban plume from the surface downwind of the city and if the plume would remain intact aloft during the night.

At 0858 on June 8 the gondola was released from the Arrowhead Airport to the west of St Louis. Staying relatively close to the mission-designed altitude of 2500 ft msl, the balloon first traveled southeastward into the southwestern portion of the St Louis metropolis, where at 1130 it turned abruptly northward, traversing the western edge of St Louis and crossing the Missouri River east of St Charles at around 1500. An hour later, the balloon began a gradual turn eastward and by approximately 1930 was on a steady eastward heading. Traveling at approximately 10 mph, the balloon continued eastward past Vandalia, Illinois, turned southeastward again at 0400 on June 9, and landed some 5 h later near the Illinois–Indiana State line.

During the daytime portion of the Da Vinci flight, similar ozone concentrations were measured from both the gondola and the mobile vans, the one exception being at approximately 1500 (Figure 9). At this time, the balloon was caught in a strong thermal updraft and was carried up above the mixing layer to an altitude of some 7000 ft, with associated decreasing ozone concentrations, before returning to its design altitude approximately an hour later.

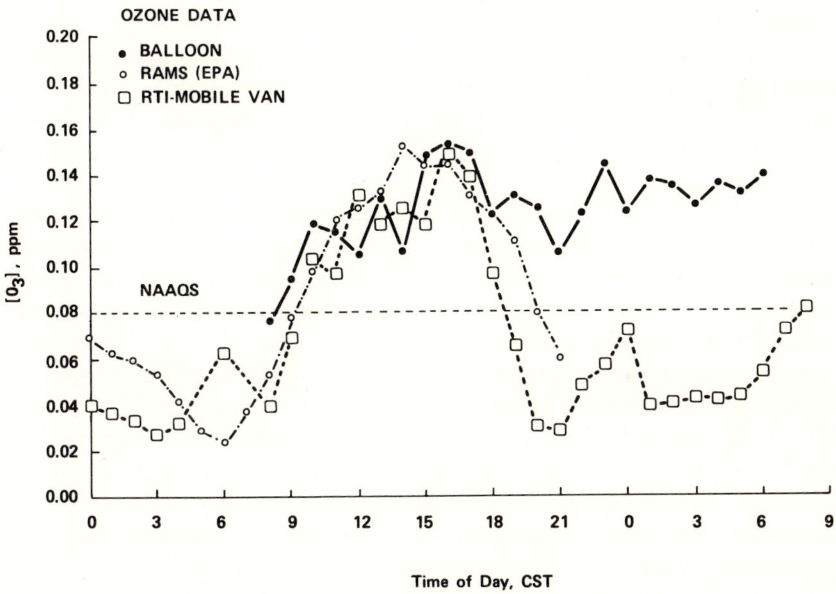

Figure 9 Ozone aloft and at ground level, June 8–9, 1976.

Significantly, the sharp drop in ozone concentration measured at the surface after sunset was not experienced aloft. The gondola, never in the center of the urban plume but perhaps at its periphery, detected less than a 20% reduction in ozone concentration after sunset, some of which can be attributed to horizontal dilution.

The measurement of ozone from the Da Vinci gondola in a quasi-Lagrangian model and other evidence show conclusively that ozone can travel great distances without significant diminution of concentration.

Conclusions

Some of the most significant conclusions of rural ozone studies are the following. Recent long-term rural background data (i.e., periods greater than 10 yr) indicate an increase in the mean rural ozone concentrations, a condition which is uncorrelated with any other identifiable geophysical parameter. Synoptic ozone relationships are identifiable with high-pressure systems; the ozone concentrations are associated with high population densities. Field data indicate that high concentrations of ozone observed at rural locations are most often generated in the lower troposphere. The air on the back side of an eastward-moving high-pressure system experiences a longer residence time than air in the leading edge of that system and, as a consequence, high ozone is generally found on the back side of high-pressure systems.

Although there may exist an area-wide system (a radius of 250 miles or more) of high ozone concentrations in which most features suggestive of a precursor origin have been smoothed out, a short-range urban influence on both hydrocarbon and ozone concentration can be observed. The evidence indicates an observable urban influence extending as far as 150 miles downwind of the city.

Transport of fossil ozone can occur for several hundred miles. The delivery of high concentrations of ozone to rural sites depends on both the movement of so-called fossil ozone and the synthesis of fresh ozone from systems in which concentrations of NO_x are low and the hydrocarbons are mainly in the form of alkanes and aromatics.

Results of recent research programs imply that the control of hydrocarbons in any individual city will reduce, but not necessarily prevent, the occurrence of high rural ozone concentrations in excess of the NAAQS at any given rural site. The implication is that the release of hydrocarbons and oxides of nitrogen from both anthropogenic and biogenic sources, located in both urban and rural areas, results in the generation of appreciable quantities of ozone over widespread regions.

REFERENCES

1. C. F. Schönbein, "Recherches sur la nature de l'odeur se manifeste dans certains actions chimiques," *C. R. Acad. Sci. Paris*, **10**, 706 (1840).
2. L. F. Evans, N. K. King, D. R. Packham and E. T. Stephens, "Ozone measurements in smoke from forest fires," *Env. Sci. Tech.*, **8**, 75 (1974).
3. C. E. Junge, *Air Chemistry and Radioactivity*, International Geophysical Series, Vol. 4 (Academic, New York, 1963) pp. 37–59.
4. V. H. Regener, "Exploring the atmosphere's first mile," in *Instrumentation and Radiation Evaluation*, Vol. 1 (Pergamon, New York, 1957) pp. 188–198.

5. W. N. Stasiuk, Jr., and P. E. Coffey, "Rural and urban ozone relationships in New York State," *J. Air Poll. Cont. Assoc.*, **24**, 564–568 (1974).
6. P. E. Coffey and W. N. Stasiuk, "Evidence of atmospheric transport of ozone in urban areas," *Env. Sci. Tech.*, **9**, 59–62 (1975).
7. G. T. Wolff, P. J. Lioy, G. D. Wight and R. E. Pasceri, "Aerial investigation of the ozone plume phenomenon," *J. Air Poll. Cont. Assoc.*, **27**, 460–463 (1977).
8. C. W. Spicer *et al.*, *Transport of Oxidant beyond Urban Areas* (Batelle, Columbus, 1976).
9. F. L. Ludwig *et al.*, *Northeast Oxidant Study* (Stanford Research Institute, International, 1976).
10. Husar *et al.* "Ozone in hazy air masses," in *Proceedings of International Conference on Photochemical Oxidant* (1976).
11. L. A. Ripperton, H. E. Jeffries and J. J. B. Worth, "Natural synthesis of ozone in the troposphere," *J. Env. Sci. Tech.*, **5**, 246–248 (1971).
12. L. A. Ripperton and J. J. B. Worth, *Chemical and Environmental Factors Affecting Ozone Concentration in the Lower Troposphere*, Final Report, National Science Foundation Grant GP-4752.
13. H. C. McKee, "Atmospheric ozone in Northern Greenland," *J. Air Poll. Cont. Assoc.*, **11**, 562 (1961).
14. J. R. Smith and C. E. Decker, *An Evaluation of Techniques for the Measurement of Low Concentrations of Trace Gases in the Atmosphere*, Final Report, National Air Pollution Control Administration Contract CPA-22-69-109 (Research Triangle Institute, 1970).
15. J. J. B. Worth, L. A. Ripperton and C. R. Berry, "Ozone variability in mountainous terrain," *J. Geophys. Res.*, **72**, 8 (1967).
16. H. L. Hamilton, Jr., J. J. B. Worth and L. A. Ripperton, *An Atmospheric Physics and Chemistry Study on Pikes Peak in Support of Pulmonary Edema Research*, Final Report, U.S. Army Research Office Contract DA-HC19-67-C-0029 (Research Triangle Institute, 1968).
17. H. G. Richter, *Special Ozone and Oxidant Measurements in the Vicinity of Mt Storm, West Virginia*, Task Report, Task No. 3, National Air Pollution Control Association Contract 70-147 (Research Triangle Institute, 1970).
18. W. S. Hering and T. R. Borden, Jr., *Ozonesonde Observations over North America*, Vols. 2–4 (Air Force Cambridge Research Laboratories, Office of Aerospace Research, 1964, 1965, 1967).
19. U.S. Environmental Protection Agency, *Control of Photochemical Oxidants—Technical Bases and Implications of Recent Findings* (Office of Air Quality Planning and Standards, Research Triangle Park, North Carolina, 1975).
20. H. B. Singh, F. L. Ludwig and W. B. Johnson, *Ozone in Clean Remote Atmospheres: Concentrations and Variabilities*, Final Report, SRI Project 5661 (Stanford Research Institute, 1977).
21. K. L. Demerjian, J. A. Kerr and J. G. Calvert, "The mechanism of photochemical smog formation," in *Advances in Environmental Science and Technology*, Vol. 4, edited by J. N. Pitts, Jr., R. L. Metcalf and A. C. Lloyd (Wiley, New York, 1974).
22. L. A. Ripperton, J. E. Sickles, II, and W. C. Eaton, *Oxidant-Precursor Relationships During Pollutant Transport Conditions—An Outdoor Smog Chamber Study*, Environmental Protection Agency Publication EPA-600/3-76-107 (Research Triangle Institute, 1976).
23. California Air Resources Board, July 1972 oxidant data (personal communication, 1972).
24. Research Triangle Institute, *Investigation of High Ozone Concentrations in the Vicinity of Garrett County, Maryland, and Preston County, West Virginia*, Environmental Protection Agency Report EPA-R4-73-019 (1973).
25. Research Triangle Institute, *Investigation of Ozone and Ozone Precursor Concentrations at Nonurban Locations in the Eastern United States, Phase I*, Environmental Protection Agency Report EPA-450/3-74-034 (1974).
26. Research Triangle Institute, *Investigation of Rural Oxidant Levels as Related to Urban Hydrocarbon Control Strategies*, Environmental Protection Agency Report EPA-450/3-75-035 (1975).

27. C. E. Decker, L. A. Ripperton, J. J. B. Worth, F. M. Vukovich, W. D. Bach, J. B. Tommerdahl, F. Smith and D. E. Wagoner, *Study of the Formation and Transport of Ambient Oxidants in the Western Gulf Coast and Northcentral and Northeast Regions of the United States*, Final Report, Environmental Protection Agency Contract 68-02-2048 (Research Triangle, 1976).
28. C. E. Decker, J. E. Sickles, II, W. D. Bach, F. M. Vukovich and J. J. B. Worth, *Project Da Vinci II: Data Analysis and Interpretation*, Final Report, Environmental Protection Agency Contract 68-02-2568 (Research Triangle Institute, 1978).

James J. B. Worth
Research Triangle Institute

Lyman A. Ripperton
Research Triangle Institute

INFLUENCE OF POLLUTION ON URBAN CLIMATOLOGY

1 Introduction

Human activity can alter the atmosphere with advertent weather-modification projects or with inadvertent upsets to the surface energy balance. An example of the former is cloud seeding to enhance precipitation, while examples of the latter include clearing forests for farming purposes, polluting the atmosphere or replacing grass and soil by concrete during urbanization. The clearing of forests on a large scale can lead to small alterations of global climate, while some investigators believe that atmospheric pollutants can do the same. The discipline of urban climatology is concerned with how urbanization and pollution produce major alterations of climate on a small scale within a city.

The period since the WMO–WHO Symposium on Urban Climate and Building Climatology held in Brussels in 1968[1] has been characterized by an exceptionally large outpouring of publications in the field of urban climate. This surge coincides with a period of increasing concern over the state of the atmosphere, especially with regard to the urban air-pollution problem.

This article summarizes how urbanization and pollution (defined as anthropogenic releases of aerosols, gases, heat and moisture) create differences between the energy and moisture budgets in adjacent urban and nonurban areas. The article also reviews the resulting urban climate as compared to that in the surrounding nonurban environs, as well as efforts to model urban climate using physicomathematical techniques.

Although average annual differences between urban and rural climatic parameters are generally small, these differences result from the redistribution of enormous amounts of energy. In addition, given correct meteorological conditions, these differences can be extremely large for many climatic variables for time periods on the scale of hours.

Finally, various aspects of the climate within cities are very important with respect both to urban air-pollution patterns and to human health and comfort. For example, research has shown that the extra warmth of an urban area increases mortality rates during regional heat waves as compared to those at nearby nonurban sites and that weak elevated urban inversions formed due to radiative loss at the top of elevated haze layers act to inhibit the vertical spread of pollutants from surface area sources. This last effect thus completes a cycle in which pollutants act to create an aspect of urban climate and then subsequently have their vertical distribution controlled by that very same urban climatic element.

There have been numerous previous surveys of urban climatology,[2–15] as well as other texts summarizing the urban climate.[16–20] Finally, in his role as Rapporteur on Urban Climatology, Oke has reviewed recent research in this field in two reports to the World Meteorological Organization.[21,22] These surveys, and in particular the two reports to the W.M.O., were essential in compiling the present review.

2 Observational Studies

2.1 *Energy Balance*

The radiation balance at air–Earth interfaces is given by

$$Q^* = (1 - A)K\downarrow - L\uparrow + L\downarrow; \tag{1}$$

the symbols are defined in the Nomenclature and each flux is positive when directed toward the surface. An example of the diurnal variation of the terms in this radiation budget is given in Figure 1.

One problem which arises in evaluating Eq. (1) in urban areas is the determination of the "surface," given the complex nature of urban structures. Fluxes in volumes between buildings differ from, but are related to, fluxes above buildings, and this article follows the terminology suggested by Oke,[23] in which the *urban canopy layer* and the *urban boundary layer* are the regions below and above roof level, respectively. The energy balance at an extensive integrated urban surface, as viewed from above roof level, will be described first, followed by descriptions of energy budgets within urban canopy and boundary layers.

The incoming global short-wave (solar) energy received at urban surfaces is about 5–15% less than that received at nearby rural surfaces on an annual basis.[22] The differences are greatest for heavily polluted air when path lengths are large at low solar elevations. The incoming solar energy is composed of direct and diffuse components, and atmospheric gases and particles reduce the intensity of direct solar energy by absorption, reflection and scattering, while increasing the proportion of diffuse energy by forward scattering to the surface.

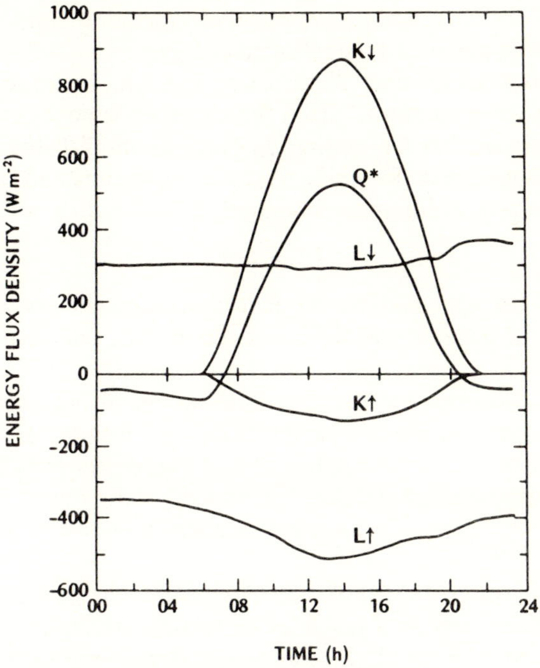

Figure 1 Diurnal variation of the terms in Eq. (1); measurements at Matador, Saskatchewan, at 50°N latitude on July 13, 1971 over a 0.2 m stand of native grass.

Observations at nearby urban and rural sites at Adelaide indicate that pollutants increase scattering more than they increase absorption.[24] In extreme cases, the direct solar beam can be reduced by more than 50% as compared with "background" levels for nearby "clean" skies with solar zenith angles of 55–65°.[25] However, the loss of direct solar energy was partially compensated by a gain in diffuse energy; the degree of compensation varied from 58% at a solar zenith angle of 65° to 70% at an angle of 37.5°.

Pollutants are most effective in scattering energy at shorter—especially UV—wavelengths. This filtering produces a condition called the urban "ultraviolet night," which has implications for skin cancer[26] and photosynthesis.[27]

The upward-directed short-wave radiation—given by the product $AK\downarrow$—is composed of upward-directed diffuse solar radiation and that reflected by urban surfaces. Aircraft observations over St Louis yielded urban and rural albedos A of 0.13 and 0.16, respectively.[28] Differences in albedo are a function of the solar elevation as a result of a combination of complex factors acting to decrease (e.g., trapping of energy between building elements and shading) and increase (e.g., removal of vegetation and the use of light-colored building materials) the reflected energy.

Aircraft observations over Zaporozhye, U.S.S.R., found greater upward-directed long-wave energy $L\uparrow$ from the city than from its environs during night-time periods, but less during day-time hours.[29] If the emissivity ε is identical for both areas, this implies warmer night-time and cooler day-time urban surface temperatures as compared to rural values, as $L\uparrow$ is given by

$$L\uparrow = \varepsilon\sigma T_0^4. \tag{2}$$

As shown in the next section, the first conclusion is very consistent with observations of urban–rural differences in surface temperature, while the second is less so.

It was demonstrated by Yap that if urban emissivities were slightly less than rural values, say 0.92 as against 0.98, then it is possible for day-time urban values of $L\uparrow$ to be smaller than rural values even though urban surface temperatures are greater.[30] Emissivity values must be determined with greater accuracy before surface temperatures can be obtained using Eq. (2).

Observations of downward-directed long-wave energy $L\downarrow$ show increased values in urban areas as a result of pollutants, which absorb energy ($K\downarrow$ and/or $L\uparrow$) and then reradiate heat with a large downward-directed component, which is included in $L\downarrow$. Night-time urban-rural differences in $L\downarrow$ appear to be small,[31] but Rouse *et al.* found large differences during day-time periods with low solar zenith angles.[32]

The all-wave surface radiant flux Q^* can be determined from simultaneous observations of each term on the right-hand side of Eq. (1), but there are few simultaneous observations at adjacent urban and rural sites. Although each term is modified by urbanization, the net effect is small, because the various changes offset one other. For example, $K\downarrow$ is attenuated, but surface short-wave absorptivity is increased, and thus net urban–rural short-wave differences are likely to be small.

Likewise, $L\downarrow$ is greater over urban areas, but the greater warmth of cities yields a larger $L\uparrow$ (at least at night), and thus net long-wave losses are only slightly greater in cities. Combining the net long- and short-wave fluxes yields a tendency for cities to have slight radiative deficits throughout the day and night in comparison with nearby nonurban areas.

Energy balances for the urban "surfaces" described for Eq. (1) may be written as

$$Q^* + Q_F = Q_H + Q_E + Q_G. \tag{3}$$

An example of the diurnal variation of the terms in this heat balance is given in Figure 2. In Eq. (3) the anthropogenic heat flux due to combustion, Q_F,

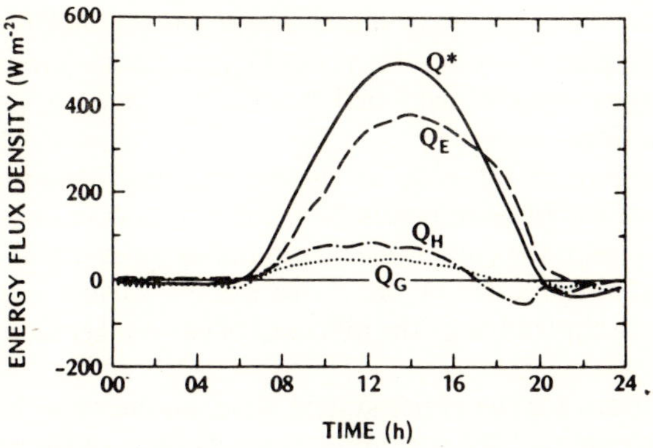

Figure 2 Diurnal variation of the terms in Eq. (3); observation at Pitt Meadows, British Columbia, at latitude 49°N on July 25, 1976, over a 0.25 m stand of irrigated grass with clear sky conditions.

represents a net energy source for the system. It is usually evaluated using fuel-consumption data and, on an annual basis, can be greater than Q^* in high-latitude, densely populated or industrialized cities such as Manhattan, New York.[33] In winter, Q_F can be the dominant energy source, while in summer it is usually insignificant in comparison with solar energy.

Over idealized rural surfaces, Q_F is zero, and the net day-time radiant surplus Q^* is divided between energy conducted into the ground Q_G, energy convected into the atmosphere Q_H and energy used for evapotranspiration Q_E. At night, the net radiant deficit is compensated by the conduction of energy to the surface through the soil, convection to the surface from the atmosphere and condensation.

There are few observations of the partitioning of energy into the terms on the right-hand side of Eq. (3). One problem involves the "subsurface" conduction term Q_G, which becomes the net rate of energy storage within the urban canopy boundary layer. This layer is very complex and consists of a soil–concrete-building–air volume.[34]

Direct measurement of Q^* and Q_H above Vancouver, Canada,[35] coupled with the use of roof heat-flux plates to approximate Q_G, left Q_E as a residual in the energy balance. Under ideal summer conditions Q_E was found to be greater than Q_H, suggesting that urban areas are not as "dry" as is generally believed, because of urban irrigation and the significant number of natural surfaces left in some cities.

The discussion so far has been concerned with fluxes of energy at roof level, but recent studies have studied fluxes within a north–south oriented urban canopy in Vancouver.[36] Albedos within the canopy were found to be 20–50% less than those of similar horizontal surfaces, as a result of increased short-wave trapping.

Although the urban geometry caused phase shifts in the diurnal pattern of Q^* for various canyon components (walls and floor), the budget for the complete canyon system exhibited reasonably smooth and symmetric temporal variations.

The partitioning of $Q^* + Q_F$ in the canyon was determined by direct measurement of the radiant, conductive and evaporative heat fluxes for the walls and floor, with the convective flux found as the residual. The temporal variations and magnitudes of the fluxes for the various surfaces differed greatly and were controlled by the influence of canyon geometry on radiative exchanges.

Diurnal patterns for the entire system were, on the other hand, relatively smooth and symmetric. By day, approximately 60% of the radiant surplus was dissipated as Q_H, 25–30% as Q_G and 10–15% as Q_E. At night, canyon winds were near calm, and the radiant deficit was almost totally balanced by heat released from subsurface storage.

The discussion thus far has centered on energy balances at roof level and within canyon boundary layers, but research is also being carried out on the effects of radiative-flux divergences due to pollutants on the thermal structure of urban boundary layers. The Complete Atmospheric Energetics Experiment (CAENEX) has studied radiative transport in the lowest kilometer of Zaporozhye.[37–39,29]

Results indicated that short-wave flux convergence in the urban planetary boundary layer due to gases and aerosols is greatest at midday, when it produces a warming of about 0.1°C/h greater than in nearby rural air. Long-wave convergence, which during some afternoon hours produced warming approaching that due to short-wave absorption, weakened in the evening and was replaced by cooling at night. The all-wave convergence produced a warming of about 0.4°C/h, of which the long-wave component contributed about 0.03°C/h. Spectral measurements showed greater convergence in the urban area at all wavelengths, and aerosols were found to absorb shorter wavelengths preferentially.

There still remains a great need for detailed observational studies on the radiative properties of aerosols and their effects on the thermal structure of the urban atmosphere.

2.2 Moisture Balance

The surface water balance of an urban area may be represented by the relation

$$P + I + Q_q = E + R + S; \qquad (4)$$

hydrologists and engineers are interested more in imported water I, runoff R and net storage S, while atmospheric scientists are interested more in precipitation P, anthropogenic moisture due to combustion Q_q and evaporation E. Urbanization and pollution act to change the relative roles of each component in Eq. (4) with respect to rural sites. Unfortunately, the two groups have not yet meshed to carry out an integrated study relating all components.

The engineering terms I, R and S are well covered in the engineering and hydrological literature, which includes several useful summaries.[40–43] Results show that the imported water supply I is increased in urban areas because of piped-in water, while the runoff R is also increased as a result of the lower permeability of urban surfaces and also of channelling. The storage S is decreased in a city, however, as a result of poor inception and infiltration.

This review is primarily concerned with the effects of cities on the precipitation (type and amount), the interception and storage of precipitation and the release of water vapor through combustion. Precipitation is generally increased as a result of the physical, thermal and mechanical uplift of air flowing over the city and also of the increased numbers of condensation nuclei. These processes will be discussed in the section on urban climate, and the subject of urban evapotranspiration was reviewed in connection with latent heat fluxes in the previous section.

Anthropogenic vapor-release rates have been calculated from fuel inventories; in particular, it has been demonstrated that Q_q in New York can contribute to an observed night-time urban excess of atmospheric moisture.[44] In Edmonton, Q_q was shown to explain about 30% of the observed winter excess in humidity at night at temperatures of about 0°C.[45] During summer days in St Louis, vapor releases have been found to be negligible by comparison with estimates of day-time potential evapotranspiration.[46]

The alterations in urban energy and moisture budgets discussed above act to create a climate within cities which differs from that at nearby rural sites.

2.3 Climatic Elements

Traditionally, studies of the effects of cities on weather elements form the largest portion of the urban-climatological literature. These studies provide considerable practical information for the fields or urban weather forecasting, urban and architectural design, heating and air-conditioning engineering, construction, air-pollution modeling, urban planning, land-use management, human comfort and health care.

2.3.1 Temperature The heat-island effect of a city (its relative warmth as compared with preurban conditions, or as it is approximated by present urban–rural temperature differences) is a central research focus in urban climatology.

The thermal modification of the canopy boundary layer is usually measured by conventional station networks or by automobile traverses. Its detailed morphology consists of two basic scales. First, there is an overall outline related to the shape of the urban region and its unique geographic features (topography, water bodies, etc.) and second, there is an internal pattern strongly controlled by microscale factors related to urban land uses and building density, such as: (1) Q_F from building sides; (2) greater absorption of $K\downarrow$ as a result of the canyon geometry; (3) decreased $L\uparrow$ as a result of the reduction of the sky-view factor by the canyon geometry; (4) greater daytime Q_G (and its nocturnal release) as a result of the thermal properties of the building materials; (5) greater Q_H as a result of the reduced Q_E due to the removal of vegetation; and (6) convergence of Q_H because of a reduction of wind speed within the canopy.

On a daily basis, the above factors retard the nocturnal cooling of urban surfaces between sunset and midnight, when rural surfaces are cooling most rapidly. After midnight, Q_F is reduced,[47] and both surfaces have reduced values of Q_G, since most of their stored solar energy has been exhausted. Thus, from about midnight until morning, both surfaces cool at about the same rate (about one-half of the early-evening rural rate).

Solar heating causes more rapid warming of rural surfaces as compared to urban surfaces, as a result of the thermal properties of urban building materials. Other factors retarding early morning warming of urban surfaces include shading by the tall buildings and increased albedo associated with elevated pollutant layers.

From the above discussion it follows that surface heat islands achieve maximum values near midnight and remain fairly constant through the remainder of the night. Heat islands then decrease after sunrise and can become weak urban "cold islands" at around midday.[48] In either case, most studies of day-time heat islands generally show small urban–rural surface-temperature differences.

The annual variation of surface heat-island values shows maximum values in summer for low-latitude cities where the storage of solar energy in urban building materials is the main factor in heat-island formation. Winter maxima are found in high-latitude cities where Q_F is the dominant process.

The magnitude of the surface heat island at a particular time and place depends on local meteorological conditions such as cloud cover and wind speed. Clear skies favor rapid nocturnal cooling of rural surfaces and hence lead to strong urban heat islands. Low wind speeds allow excess urban heat to accumulate, and thus the normally high day-time wind speeds (and their associated high turbulence levels) are also important reasons why daytime heat islands are generally weak.

The average annual value of the maximum surface urban heat island ΔT_0

has been related to population using the formula[49]

$$\Delta T_0 = 0.25 \frac{P^{1/4}}{u^{1/2}}. \tag{4}$$

It yields a value of about 2.5°C for cities with populations of 1000 and about 10.2°C for cities with one million persons. However, heat islands have been found around refinery complexes, shopping centers[50] and even single buildings.

The thermal modification of the urban boundary layer has been studied using tall towers, helicopters and fixed-wing aircraft. Results generally show that during night-time periods with rural surface-based radiation inversions, the urban boundary layer from the surface to several hundred meters is neutral or slightly unstable with one or more weak, thin (100-m), elevated inversion layers aloft.[33] The urban elevated inversion layer is believed to form, partially at least, from radiative-flux divergence at the top of the urban haze layer.

The vertical variation of heat-island intensity at night generally shows decreasing magnitudes with height over the layer extending to the top of the urban elevated inversion. Above this layer there is often a weak "crossover layer," in which urban temperatures are lower than rural values at the same level. This layer is several hundred meters in depth and arises from radiative cooling of the haze layer and/or urban convection cells with compensating subsidence over peripheral open lands.[51]

The shape of the nocturnal urban boundary layer tends to be dome-like during near-calm conditions,[33] and plume-like during higher-speed periods,[52] when the layer downwind of the city is found to lie above the normal rural stable boundary layer.

During the day, when the rural surface boundary layer—the lowest few tens of meters of the planetary boundary layer (PBL)—is unstable and the remaining rural PBL is close to neutral, the entire urban PBL is close to neutral, because of the increased mechanical turbulence associated with rougher urban surfaces. However, the increased mechanical turbulence, as well as any thermal turbulence associated with the normally weak day-time heat islands, leads to increased convective activity, which can "dome" regional inversions located one or two kilometers above the surface.[53]

2.3.2 Wind Winds in the atmospheric boundary layer are governed by a balance of the pressure-gradient force, the frictional stress and the rotation of the earth (*via* the Coriolis force). Air traversing an urban area encounters changes in surface and atmospheric characteristics which alter the existing balance. In particular, urban heat islands impose three-dimensional deformations of the pressure field, and rougher surfaces increase the frictional drag exerted on the airflow. These changes (and others) combine to modify the horizontal and vertical wind velocity (speed and direction) for the mean and turbulent components of the flow.

The roughness of a surface is parameterized by the surface-roughness length z_0, the value of which can be determined from wind observations at several levels within the urban planetary boundary layer. Typical values include 0.01–0.2 m for open countryside, 1.0–1.5 m for woodlands, suburbs and small towns and 2–4 m for the centers of large cities.

The roughness of urban surfaces and the decreased stability of night-time urban atmospheres increase (by about 30–50 %) the intensity of mechanical and thermal turbulence over a city.[54] In addition, the peak in the energy-spectra curves shifts toward higher frequencies for a city.

Urban areas tend to decelerate regional flow speeds during high-speed periods and to accelerate them during low-speed periods. The deceleration is due to increased surface frictional drag, and the acceleration to pressure gradients associated with urban heat islands, which are strongest during low-speed conditions. The "critical wind speed" for the change from deceleration to acceleration is 4.5 m/s in London,[55] and 3.6 m/s in New York.[56]

Changes in speed extend to several hundred meters in the urban boundary layer and are capped by "return-flow" layers of equal depth, in which the reverse effect is found.[57] Return-flow layers are similar to those found above sea-breeze flows, and are most noticeable during otherwise calm conditions when warm urban areas generate their own sea-breeze-like circulation systems. Such circulations in St Louis were found to extend to a depth of 1 km and to have a surface inflow velocity of 2 m/s, an updraft vertical velocity of 0.3 m/s and a lifetime of 1–4 h.[58]

Changes in the flow speeds of air over urban areas lead to adjustments in the balance of forces and thus to changes in wind directions. When wind speeds decrease, the Coriolis force decreases and, since the pressure-gradient force may be assumed constant, the flow must turn towards lower pressure (i.e., cyclonically).

Heat-island-induced circulations cause a streamline convergence of air near the surface into the center of the city from all directions. Above this layer there is an outflow region of divergence. During periods when the prevailing flow is accelerated by heat islands, the convergence region is superimposed on the existing flow field, and is advected downwind from the center of the city.[56]

The magnitude of convergence over St Louis was estimated to be about $(5–10) \times 10^{-5}$ m/s,[59] which leads to a mean uplift of about 0.1 m/s. Other factors causing upward motion over cities include the barrier effect of the physical presence of the city and deceleration of the flow due to frictional retardation. Factors causing downward motion over a city include the acceleration of existing flows by the heat island.

A study of slow-moving synoptic cold fronts in New York City found that, during periods with no heat island, these fronts were retarded by at least 50% over the central urban area as compared with upstream speeds; this was the result of increased surface drag.[60] Conversely, during periods with well-developed heat islands, fronts were retarded over the upwind half of the city as before, but were accelerated by at least 25% over the downwind half of the city, as a result of heat-island pressure fields. In a related study, sea-breeze fronts were also found to be retarded over the center of New York.[61]

2.3.3 Humidity Analysis of urban–rural surface dew-point data for Chicago showed seasonal and diurnal variability in urban–rural moisture differences.[62] In winter, the city was moister during day-time hours as a result of anthropogenic moisture production, Q_q, while at night it was moister because of both Q_q and reduced urban dewfall (condensation or a negative E). In summer (when Q_q is small), the urban area was moister only at night, while during day-time periods it was drier as a result of reduced urban evapotranspiration rates E. Data from Edmonton showed a winter urban excess of moisture, which increased as the city grew, and thus as Q_q increased.[45]

Results from helicopter soundings in the urban boundary layer are, for the most part, consistent with the above results; for example, New York was found to have an urban "vapor dome" in the lowest 500 m of the atmosphere, in which absolute humidities were greater than rural values.[63] The early-morning moisture excesses were 0.9 g/m^3, while the afternoon excesses were only 0.2 g/m^3. Similarly, specific humidity deficits of about 1 gm/kg were found over St Louis during summer midday periods.[64]

Although absolute humidities in urban areas are higher than in rural areas during periods when Q_q is a significant source of moisture, the higher temperatures of the urban heat island create lower urban relative humidities during these periods in both the canopy layer[65] and the boundary layer.[63] In cloud physics and physiological studies it is often the relative humidity—i.e., the degree of saturation—which is the important factor and not the actual moisture concentration.

2.3.4 Clouds and fog Urban areas have increased levels of clouds and fog as compared with nearby rural areas even though, as discussed above, urban relative humidities are less than nearby rural areas. As an overall average, large cities have 25% less clear days and 5–10% more cloudy days than their rural environs.[66]

One factor increasing the amounts of cloud and fog in urban areas is the large number of giant ($1 < r < 10\ \mu$m) hygroscopic nuclei which both attract water, allowing condensation at subsaturation relative humidities, and enhance cloud-droplet coalescence processes. A second factor is increased upward motion over urban areas, which also leads to higher cloud bases and greater cloud instability. This motion is due to the barrier effect of urban structures, mechanical and thermal turbulence and also convergences in the flow field.

Much new information about urban effects on clouds and precipitation has come from the Metropolitan Meteorological Experiment (METROMEX) being carried out in and around St Louis. While the city was found to be a major source of both Aitken condensation nuclei ($4 \times 10^{-3} < r < 0.1\ \mu m$)[67] and total cloud-condensation nuclei by Auer (1975),[68] urban pollutants may actually poison ice-forming nuclei.[69] Gas-to-particle transformations involving existing nuclei may also play an important role in the growth of particulate mass in urban plumes.[70]

St Louis, together with certain industrial areas, appears to be a preferential site for the initiation of small cumulus clouds,[71] and convective cloud bases are generally higher by 300–600 m over the city.[72] Urban effluents ingested by clouds, as well as enhanced cloud-condensation-nuclei levels, result in clouds over St Louis with more uniform droplet sizes than those over rural areas.[73] Urban aerosols and urban-induced changes in cloud structure have been found up to 80 km downwind of the city.[74]

Radar echoes and precipitation cells in summer cumulus clouds over St Louis show rapid initiation and urban enhancement of warm rain processes, probably as a result of giant nuclei from urban sources.[75] The city's heat island also initiates local convection, clouds and rainfall,[76] while storms over St Louis produces more large rain drops, and larger and more numerous hailstones.[77]

A 10–30% increase in summer rainfall downwind of St Louis produces an average increase of 15% in river runoff,[78] as well as greater infiltration of groundwater. Tracer materials released at the surface are taken into the clouds and appear in subsequent rain areas 5–35 km downwind of the city,[79] while groundwater contamination has increased rapidly, partly as a result of the deposition of pollutants scavenged from the St Louis urban plume.[78] Annual average crop yields, as well as crop losses due to hail, have both increased downwind of the city.[78]

Cloud studies over other cities have obtained results consistent with those from METROMEX.[80,81] Satellite images from ERTS have shown cumulus bands being spawned downwind of Chicago,[82] and downwind of industrial sources in association with large point-source plumes.[83]

2.3.5 Visibility The prevailing visibility is defined as the greatest visibility attained or surpassed throughout a half—not necessarily continuous—of the horizon circle. The visibility is reduced by contrast reduction resulting when fog and/or haze particles affect that portion of the solar beam reflected off large objects. As haze particles (0.1 μm) are smaller than fog droplets (10 μm), they differentially scatter the shorter wavelengths of solar energy into the reflected beam. This produces a bluish color when the haze is viewed against a dark background, as only the slightly bluish scattered light reaches the eye. Fog droplets reflect equally all wavelengths in the beam reflected from an object, and hence create a gray sky.

Correlation of visibility with the number of days with fog and/or haze in New York City from 1950 to 1960 showed that an increase of days with high levels of atmospheric pollution caused an increase in the number of days with low visibilities of 4–6 miles.[84] Similar results were obtained in Sydney.[85]

3 Theoretical Studies

3.1 Canopy Layer

Models of the energy and mass exchanges inside typical urban canyons have been developed, with the radiative exchanges most amenable to analysis. A model of the absorption of short-wave radiation by buildings and streets has been developed by Terjung and Louie.[86] Their methodology involved the definition of a "radiation neighborhood" (a strip extending from the top of one building, along the canyon walls and floor, to the top of the building on the other side of the street) whose form is repeated throughout the urban area, but whose dimensions (such as height and width) are kept flexible. Using view-factor algebra and multiple integration, the direct- and diffuse-beam short-wave radiation received and reflected by components of the "neighborhood" were obtained. The general model was then applied to a hypothetical "synthetic city" of given shape, structure and dimensions, for different latitudes and seasons.

In all cases it was shown that the central areas of the city absorbed more strongly than locations on a surrounding horizontal plain entirely as a result of changes in surface geometry. The output also suggested that by day the maximum heating due to solar absorption will occur around the margins of the downtown area, and not in the center itself, as a result of shading by tall structures. This conforms with the negative heat islands found at midday in cities with very tall buildings in their core. Although the geometry may result in much-enhanced absorption by building systems, this may not necessarily apply at street level and therefore observations near the ground in canyons could lead to erroneous conclusions.

This model was subsequently modified to permit the use of average view factors for sunlit and shaded areas, with new submodels to compute long-wave, heat-storage and turbulent-transport terms.[87] These latter exchanges were related to an equilibrium surface temperature, or to empirical relations derived in other studies. Evaporation from wet surfaces was calculated using the Penman–Monteith combination model with assigned aerodynamic and surface resistances.

An analogous approach was undertaken by Arnfield,[88] who modeled the albedo of a canyon, including the effects of multiple reflection. Various height-to-width ratios, solar geometries and albedo mixes were modeled, and the approach allows maps of surface radiative properties to be constructed for any city from urban land-use maps and surveys. He also developed a second model for calculating the long-wave emissivity of a canyon, although this was not tested. Much neglected in the past, long-wave radiation modeling for nonhorizontal surfaces is now gaining impetus, especially as a result of the work of Unsworth[89] and Cole.[90,91]

A model of solar-radiation exchanges in an experimental canyon, using parametrizations to obtain the net all-wave radiation transfer Q^*, the sub-surface-storage transfer Q_G and the turbulent sensible heat transfer Q_H, yielded good agreement with direct measurements for the solar-radiation input $K\downarrow$ and the absorption $AK\downarrow$.[92] It was found possible to relate $K\downarrow$ to Q^* and Q^* to Q_G, and thereby to solve for Q_H in dry conditions where latent heat could be neglected.

The vortex circulations that form in canyons when the air flow is perpendicular to their axial orientation, and the jetting flow when winds are parallel to this axis, have been modeled on the assumption that the characteristic general wind profile for the urban area consisted of a logarithm section above roof level, and an exponential section in the obstructed (canopy) layer beneath. Vertical and horizontal transport equations were then incorporated into a model to predict volume-average concentrations of carbon monoxide from automobile sources in the canyon.[93]

Much of our knowledge of airflow at the building scale has been derived from wind-tunnel studies using scale models. The work is usually concerned with the effects of an individual (usually large) building in an urban context.[94–96]

3.2 Boundary Layer

The review of numerical models at the urban scale is broken down into four categories: energy-balance models (one-dimensional); time-dependent advective integral models (two-dimensional); dynamic differential models (two-and three-dimensional, time-dependent); and scale models. Some of the outstanding problems in this field have been reviewed elsewhere.[97–99]

3.2.1 Energy-balance models One-dimensional time-dependent models of the atmospheric boundary layer are mainly used to study the effects of pollutant concentrations on the thermal structure of the atmosphere, and its feedback to atmospheric stability and pollution dispersion. The most complete models consists of a subsurface layer, an analytical constant-flux surface boundary layer and an upper numerical planetary boundary (or transition) layer. In most cases, the following assumptions are made for the upper layer: (1) that the fluid is hydrostatic; (2) that the flow is incompressible; and (3) that horizontal advection is more important than horizontal diffusion.

The models predict the time and height variations of atmospheric wind, temperature and pollution concentration, as well as soil temperature, the latter from the one-dimensional molecular heat-conduction equation. In the constant-flux layer (about 50 m in depth), the vertical fluxes of momentum, heat and pollution are assumed to be constant with height but not with time, and the vertical variations of wind, temperature and concentration are obtained from analytical expressions dependent on the time-varying atmospheric stability.

The time and height variations of wind, temperature and concentration in the planetary boundary layer are obtained from numerical solutions to the finite-difference analogs of the momentum, energy and pollutant-concentration equations, respectively. These equations each contain a vertical eddy-diffusion term, the following processes also being included: rotation of the earth (Coriolis force) and pressure-gradient force in the momentum equation; and radiation-flux divergence in the energy equation.

Boundary conditions at the bottom of the soil layer and the top of the PBL must be specified, as well as expressions for the eddy-mixing coefficients. Surface winds are assumed to be zero, and the magnitude of anthropogenic fluxes must be given. The surface temperature is computed by using iterative techniques to obtain solutions to Eq. (1) and (3), where fluxes in the latter equation represent fluxes in the surface boundary layer.

Solutions over urban areas were first obtained by Tag,[100] who did not include the effects of radiative-flux divergence and those of Q_G in his surface energy equation. The only result that he presented was the time variation of the surface heat island, and this parameter was predicted to be negative during early morning hours, as the thermal properties of the urban surfaces overcame the lower urban albedo and reduced evaporation.

A similar study was carried out by Bergstrom and Viskanta,[101,102] except that radiative flux divergence due to gaseous and solid pollutants in the atmosphere was included for the solar portion of the spectrum, the transfer equation being solved using a spherical-harmonics approximation.[103,104] The long-wave divergence was predicted using the observed water-vapor and carbon-dioxide emissivity data. Model computations were compared with rural observations before conducting a series of simulations of the thermal structure of an urban atmosphere for summer and winter conditions.

The results suggested that pollutants would lead to a decrease of solar radiation reaching the surface, resulting in a lower surface temperature but a higher atmospheric temperature. They also increased the long-wave flux to the surface from the atmosphere, which therefore tended to offset the solar effects.

More recently Venkatram and Viskanta simplified the solar-radiation submodel (to save computer time) and greatly expanded the turbulence portion.[105–107] The solar fluxes were computed according to the two-stream method, and the results showed that the predominant influence of pollutants was to produce surface warming despite the reduction in the surface receipt of short-wave radiation. Note that this result is the reverse of that predicted by the earlier model, illustrating the importance of accurately incorporating the role of the surface-energy balance (especially evaporation) and turbulence. In the atmosphere the thermal effects of pollutants led to an increase of stability by day, and a decrease at night.

A similar model was constructed by Ackerman to study the effects of aerosols on urban climate.[108] It included a four-stream radiative-transfer submodel based on earlier work,[109] which treated both solar and long-wave fluxes, and a boundary-layer heat-transfer model incorporating a surface-energy balance equation, and a variable mixing depth. The energy balance, however, contained only an assumed value of the Bowen ratio and no anthropogenic heat. The model was tested against the same rural data. The agreement between the model and measured radiative and temperature characteristics was considered reasonable, but sensible-heat-flux comparisons were rather poor. The model was then initialized using urban data from Los Angeles and sensitivity tests showed that the model was most sensitive to surface properties (especially albedo and Bowen ratio). The results strongly suggest that the surface–atmosphere system is characterized by self-compensating feedbacks, which combine to reduce the impact of pollutants on climate. For example, if radiation is absorbed by pollutants it reduces the sensible heat transfer from the surface, thereby tending to keep boundary-layer temperatures about the same.

Tests of a modified form of the one-dimensional model of Pandolfo and Wickam[110] against data from Jodhpur and Toronto suggested a marked reduction of surface temperature at midday in a polluted atmosphere, minor changes in minimum temperatures and major changes in stability.[111,112]

3.2.2 Advective integral models A pioneering steady-state mixing-layer model of the urban boundary layer was formulated by Summers.[113] This purely thermodynamic model postulated the development of an adiabatic mixed layer of increasing depth as stable nocturnal air of rural origin passes over a city and is altered by turbulence. The growth of the mixed layer was related to the constant rural lapse rate, anthropogenic heat input and wind speed. Summers lacked sufficiently accurate data to test the model fully, but gained reasonable agreement with urban tower data. Pasquill modified the model to allow for general wind and temperature profiles, but this necessitates the evaluation of additional empirical constants.[114]

The addition of heat sinks to Summer's model extended its use to areas downwind of the city.[115] The energy balance of Eqs. (1) and (3) was applied to both a reference rural site and an urban site. When the model was used to predict the spatial variation of mixing-layer depth and surface temperature, on the basis of fuel-consumption data for New York City, excellent correlation was obtained between observed and predicted mixing-depth values for five nonsummer mornings.[116]

Summers' basic model also yielded reasonable agreement with observations in a study of the mixing-layer depth and temperature increase downwind of Sydney, Australia in winter; sensitivity analyses were performed.[117]

3.2.3 Dynamic differential models Before two- and three-dimensional finite-difference analogs to the dynamical equations were developed, simpler models of the urban boundary layer were tried. In Gold's model, for example, circulations similar to sea breezes developed into an urban area in otherwise calm conditions as a result of prescribed temperature differences.[118] These differences decreased with distance from the urban–rural boundary, and were assumed to be constant to a height of 91 m. It was further assumed that there were no obstructions to the path of the moving air.

When the model was used to compare computed urban breezes with observed circulations in Toronto it was found that the observed values of the urban breeze normal to the heat-island boundary were less than the predicted values.[119] This is reasonable, since the model did not include surface roughness.

Urban circulations that develop in otherwise calm conditions as a result of horizontal differences in surface heating were studied by Delage and Taylar,[120] who extended a one-dimensional finite-difference PBL model to two dimensions. All such extensions require the addition of advective terms to each of the basic equations. The soil layer was omitted, and thus surface temperatures had to be specified. In addition, the constant-flux layer was also omitted, since profile laws are not valid in near calm conditions. The prescribed differential surface cooling beneath an initially stable atmosphere produced a symmetric double-cell urban-circulation pattern and an unstable layer capped by an elevated inversion over the city.

Analytical solutions for the two-dimensional stationary linearized momentum equations were obtained by Vukovich.[121] The model ignored the Coriolis effect, and a constant linear friction was used. The results showed a rather weak two-cell urban circulation system, in which the depth and intensity of the circulation depended on the assumed stability of the boundary layer. A mean wind produced a displacement of the cells downstream, to a location which was a function of the prescribed urban heating rate and of advection by the mean wind.

A similar linear model developed by Olfe and Lee[122] investigated the coupled velocity and temperature perturbations induced by an urban heat island on a uniform flow of constant stability. The resulting urban temperature profiles from the basic model (no eddy viscosity and no Coriolis force) were described as the superposition of an eddy conduction profile and a gravity wave, which produced an elevated inversion and a crossover layer. Additional simulations, within the linearized framework, each included one of the following: (a) a three-dimensional circular heat island; (b) a two-layer atmosphere having a change in stability at a suitable altitude; (c) a constant eddy viscosity for the perturbed flow; or (d) the Coriolis force.

Nonlinear finite-difference two-dimensional numerical models have also been developed for the flow over an urban area. The first was the one-layer model developed by Meroney and Yamada,[123–125] which was the numerical analog of their wind-tunnel model discussed below. They assumed constant eddy-mixing coefficients and solved a vorticity equation derived from the momentum equation on the assumption that the boundary layer is Boussinesq. The model was used to study the mutual interactions between two heated islands and their effects on pollutant concentrations.[126]

The model of McElroy, a two-dimensional steady-state model developed from the one-dimensional model of Estoque,[127] included a constant-flux layer and a transition layer in the boundary layer.[128–131] Nonpollutant radiative effects were modeled, and diffusion coefficients (assumed equal for heat, water vapor and momentum) were computed in the constant-flux layer and then allowed to decrease to zero at the top of the transition layer (set at 2.05 km). Subsurface thermal characteristics were specified as boundary conditions for inclusion in a surface-energy balance equation, which was used to compute the surface temperature. Information on surface roughness, relative humidity, thermal properties and anthropogenic heat were specified at positions along an urban–rural cross section. Boundary conditions were also set for the upwind rural area.

When applied to the flow over Columbus, Ohio,[129] no results were presented for the wind field, but observed and simulated depths of the mixed layer were in good agreement. This model was also used to estimate the effects of alternative land-use strategies on nocturnal urban thermal structure.[130]

A similar model developed by Wanger and Yu did include vertical motions and the soil layer, but did not include a constant-flux surface boundary layer, surface moisture nor radiative effects from water vapor.[132,133] The surface temperature was predicted from a surface-energy balance equation which did not include a combustion-heating term. A variable grid was used in the vertical, while equal grid spacing was used in the horizontal. Simulations with geostrophic wind speeds of 8–15 m/s showed the heat island to be inversely proportional to speed. The results indicate that a portion of the heat island can be produced by increased roughness, and that the addition of anthropogenic releases can be significant. However, the solution would not converge for winds of greater than 8 m/s, and hence only urban deceleration was predicted.

The nonsteady two-dimensional model of Bornstein also solved the vorticity and stream-function equations, with the following exceptions.[134–136] Eddy-mixing coefficients were a function of height; the Coriolis force was included; and the vorticity was given in a form consistent with the assumption of hydrostatic equilibrium. The surface temperature was specified using the observed urban and rural cooling rates of Oke and East[137]; the formulation of the surface boundary layer followed that of Pandolfo *et al.*[138] but the radiative effects of pollutants and water vapor were omitted.

Results for flows over rough cities showed reduced urban speeds at heights up to several hundred meters as compared to values at corresponding heights at the upwind rural boundary. In addition, the surface wind turned toward low pressure as the flow passed over the rough city, but returned to the original upwind value some distance downwind of the city. Increased urban wind speeds resulted from simulation of flows over warm cities.

Simulations by the model for flows over rough, warm cities reproduced most of the features usually associated with observed urban boundary layers. The predicted wind fields consisted of patterns showing areas in which urban speeds were increased, as well as decreased, over those at the upwind rural boundary. This is in agreement with the tetroon-derived urban wind field over Columbus, Ohio, as determined by Angell *et al.*[139] In addition, the model predicted maximum windspeeds near the surface at the downwind edge of the city as observed by Bornstein *et al.*[63] It also verified the existence of a critical wind speed, as observed at the upwind rural boundary[55,63]; below this value, urban speeds near the surface are increased, while above it they are decreased. However, predicted wind-speed differences were only about 20% of the observed values. Advection of cool rural air over the warm city produced an almost adiabatic mixing layer and the simulated temperature profiles agreed well with data from New York and Montreal.

More recent versions of URBMET have included the effects of anthropogenic moisture[44] and anthropogenic heat.[140] Anthropogenic moisture was found to increase the buoyancy of the urban boundary layer and hence to increase the intensity of urban circulations, while anthropogenic heat was found to prevent the urban boundary layer from becoming stable except for a few hours after midnight.

A two-dimensional model which included water-vapor effects on radiation and a surface-energy balance equation with an anthropogenic-heat term and an evaporative flux depending on surface moisture was developed by Gutman and Torrance.[141,142] They investigated cases of winter-time nocturnal heat islands and made direct comparisons against observations from Montreal. Agreement at the surface and aloft was good, and the thermal structure of the heat island was found to be more sensitive to anthropogenic-heat changes than to surface-roughness changes. Finally, Lee and Olfe used nonlinear two-dimensional calculations to investigate temperature and velocity changes induced in air flowing over a heat island.[143] The results simulated the destruction of a nocturnal inversion and the "crossover" effect above the city.

Major developments have occurred in modeling the mixing height, especially its time development during day-time as a result of the convergence of sensible heat in the boundary layer. Heat is thought to converge both from below—as a result of the normal upward heat flux from the radiatively heated surface—and from above—as a result of "penetrative convection" (entrainment of heat from the overlying inversion). Most of this work has been over rural terrain, but Barnum and Rao have incorporated these ideas into a two-dimensional model for urban mixing heights.[144,145] With the aid of this model and various analytical techniques, the dependence of mixing depth on such factors as horizontal advection, penetrative convection and lapse rate in the stable layer was delineated and compared with observations from St Louis during selected summer days.

Three-dimensional urban heat-island models have also been formulated. Black *et al.* developed a steady-state analytical model of convective fields over heated areas in the presence of an imposed wind.[146] The Coriolis force is neglected, and the eddy-exchange coefficients are equal and a function of height only. The results predict the formation of buoyant vortex rings along axes parallel to the prevailing wind, which illustrates the importance of using three-dimensional treatments in studying convective phenomena.

The finite-difference model of Atwater is an expansion of the model of Pandolfo *et al.*,[147] and includes radiative effects from gaseous and solid pollutants, as well as from water vapor.[148–151] The model is based on Eulerian conservation equations for momentum, heat, water vapor and pollutants in the atmosphere, for heat in the soil and for momentum, heat and salinity in a water body. The model extends from below the surface to a height of about 3 km. Finite-difference solutions are defined for a three-dimensional space grid at a given time step, while interface temperatures are derived from iterative solutions to a surface heat-balance equation which included an anthropogenic heat term.

The model was used to simulate conditions in an arctic city surrounded by tundra but without pollutants,[152] and for a hypothetical midlatitude city in summer and winter.[151] In the arctic case, the results were not dissimilar to those found in the midlatitudes; the major difference was that arctic summer heat islands are more strongly negative by day. In the more general case, inclusion of pollutants had only minor effects on heat-island intensity, and on the vertical thermal structure. The major thermal change was an increase in stability. The model predictions for the vertical distribution of urban–rural temperature differences agreed will with observations in New York and Montreal.

A three-dimensional primitive equation model was applied to the St Louis urban heat island by Vukovich *et al.*[153] The model used basic forcing functions to characterize the city, including spatial distributions of topography, building height, surface roughness and surface temperature. These features were entered on a 144 × 144 km nested grid of the city and its surrounding area, with eight vertical levels to a height of 4 km. Heat and momentum fluxes were determined through simultaneous solution of the boundary-layer profile equations using "friction" or "scaling" quantities for velocity and temperature which are proportional to the surface momentum and heat fluxes. Vertical eddy diffusivities were calculated using mixing lengths related to stability functions, and the initial wind and temperature conditions were specified.

The influences of stable atmospheres were investigated using different synoptic wind speeds and directions. It was found that the heat-island intensity and its circulation decreased with increasing wind speed and shear, and that its heat plume also extended further downwind. The thermal anomaly in the weakest wind case was centered on the city and extended to about 300 m in the vertical. Similarly, the wind anomaly also was centered over the city with a pronounced zone of convergence and cyclonic curvature. Strongest winds were found upwind of the city center (where the heat-island pressure gradient produces acceleration), and weaker winds were found downwind. At 300 m there was strong divergence, and the city center was characterized by upward vertical velocities extending to about 1 km.

In simulations with stronger winds the heat-island circulation changed from one with cyclonic rotation to one characterized by a convergence line. Just as the heat plume extended downwind at higher speeds, so did a tongue of positive vertical motion. Changes in the heat island due to wind direction were largely attributed to the effects of topography.

3.2.4 Scale models An early approach to modeling the urban boundary layer was the outdoor 1 : 1000 scale model used to simulate flow over Fort Wayne, Indiana.[154,155] Hot wires and roughness elements were embedded in the ground, and measured temperatures at heights of up to 1.5 m within the model were compared, by scaling with the Monin–Obukhov parameters, to observations at heights of up to 300 m obtained from a tower in Fort Wayne. Moderate success was claimed in reproducing the stability variation across the urban complex.

A method of simulating the characteristics of adiabatic urban boundary layers in wind tunnels was discussed by Cook,[156] who carried the methodology through and applied it to a specific site for wind-loading purposes,[157] obtaining reasonable agreement with observations for such properties as the mean velocity profile, turbulence intensity, stress and gust velocity. Similarly, Counihan reported simulations of adiabatic flow over urban terrain in which reasonable agreement with full-scale data was achieved.[158] The characteristics studied included the profiles of mean velocity and turbulence intensity, power-spectral density, cospectra of stresses and the power-law index.

A wind-tunnel model of an urban area has been used by Yamada and Meroney to study the flow of stratified layers over an urban area.[123–126] The depth of the model atmosphere was about 15 cm, and its horizontal extent about 50 cm. Heat was supplied to the flow by electric heaters at the entrance to the tunnel and along its ceiling. Results of flow over heated obstacles showed changes in the flow pattern, frequent elevated inversions, less frequent surface inversions and crossover layers forming over the "city" during night-time hours.

The Colorado State University wind tunnel allows stratified boundary layers to be simulated.[159–163] In a number of experiments stable flow across a heated "city" was investigated. The heat island was simulated by the use of electrical heaters, and in some cases roughness elements (aluminum blocks) were arranged on the heated area in a street-block pattern. Flow was visualized with the aid of a passive smoke source; mean and fluctuating temperatures were measured with an array of thermocouples; and in some experiments the mean wind speed and the longitudinal component of the turbulence were monitored with temperature-compensated hot-wire probes.

The results suggested that the heat island was characterized by two large circulation cells, one upwind and one downwind. The former was associated with downwind motion near the upwind edge, and the latter with upward motion extending over about one-fifth of the heat island and downwind. It was thought that these cells would be symmetric about the heat-island center if general flow was absent. There was low-level horizontal convergence, uplift over the "city" and upper-level horizontal divergence. The reverse situation applied in areas surrounding the heat island. Longitudinal turbulence intensities were greatly enhanced over the rough heat island and showed a minimum along the sides of the "city," where lateral convergence produced longitudinal roll vortices. An elevated heat plume was seen downwind, and it was suggested that buoyancy forces caused larger perturbations in the flow than the greater roughness. An empirical relationship to predict the height of the modified layer was devised and tested.

4 Conclusions

This article has summarized the role of pollution and urbanization in producing the phenomenon known as urban climate. The various aspects of this climate have been described, as well as attempts to simulate this climate using numerical models.

Additional observations are required so that modeling will not exceed both the database and our physical insight into the workings of the urban-atmosphere system. Areas requiring additional observations and scrutiny include the absorption and scattering coefficients of aerosols, the urban albedo, the urban outgoing long-wave energy flux, interaction between radiative-flux divergence and atmospheric temperature change, the partitioning of the net radiant flux into atmospheric and subsurface heat fluxes, urban evapotranspiration, synthesis of the observations of urban precipitation patterns and the interactions between the urban canopy and boundary layers.

In the area of modeling there are various problems of concern not only to urban modelers but to all boundary-layer modelers, such as those associated with initialization and verification procedures. Urban modelers have additional problems concerning the choice of optical properties of pollutants, surface properties of urban areas, parameterization of surface evaporation, subsurface storage and recognition of the role of the canopy layer. Until these problems are solved, modelers must be careful in predicting the effects of varying a particular term.

The ideal future boundary-layer model might have the following components not yet included in any urban models: upper boundary conditions obtained from real-time synoptic forecasts; lower boundary conditions for moisture obtained from solutions to the balance equation for moisture, including time-dependent anthropogenic moisture terms; and eddy-diffusion coefficients derived from solutions to the turbulent-energy equation.

This model would then be coupled, *via* a radiative flux divergence term in the energy equation, to a similar finite-difference mesoscale air-pollution transport and diffusion model, which included photochemical transformations and other natural removal processes. The system of models could then be applied to various mesoscale urban–rural regions to solve a variety of problems including regional planning and site selection for new pollution sources.

Even now information on urban climatic effects can be incorporated into the design plans for new cities. However, the available information is only slowly seeping through to planners, as a result of the lack of the right kind of information and poor communication. In the future, more attention will be necessary in order to achieve a genuine interfacing of planners and atmospheric scientists.

Perhaps some day people will, as Dr William Lowry says, "live in cities whose festering sores are gone, the fundamental problems of human well-being solved. In the cities of that day, men can live in unstressed ecosystems, with time to attend to the finer points of adaptive urban design. Toward that end, and with the same faith as the fire-fighting forester who studies and plants trees never to be harvested in his lifetime, I urge you to recognize and respond to the short-term needs of urban man, at the same time seeing in perspective the possibilities for a scientific–humanistic growth and application of an urban climatology."

Nomenclature

A	Urban albedo, dimensionless
E	Rate of evaporation,
I	Imported water,
$K\downarrow$	Incoming short-wave radiation
$K\uparrow$	Upward reflected short-wave radiation
$L\uparrow$	Long-wave radiation emitted by the surface
$L\downarrow$	Long-wave radiation from the atmosphere
Q^*	Net all-wave radiation
Q_E	Latent heat flux,
Q_F	Anthropogenic heat flux,
Q_G	Sub-surface heat flux,
Q_H	Sensible heat flux,
Q_q	Anthropogenic moisture flux
P	Precipitation
R	Runoff
S	Net storage of water
T_0	Surface temperature
p	population
u	Wind speed
z_0	Surface roughness parameter
ΔT_0	Surface heat island magnitude
σ	Stefan-Boltzman constant
ε	Surface emissivity

REFERENCES

1. World Meteorological Organization, *Urban Climates*, W.M.O. Technical Note No. 108, W.M.O. No. 254 TP 141 (W.M.O., Geneva, 1970).

2. J. T. Petersen, *The Climate of Cities: A Survey of Recent Literature*, Publication No. AP-59 (National Air Pollution Control Administration, 1969).
3. T. J. Chandler, *Selected Bibliography on Urban Climate*, W.M.O. No. 276 TP 155 (W.M.O., Geneva, 1970).
4. T. J. Chandler, *Urban Climatology and Its Relevance to Urban Design*, W.M.O. Technical Note No. 149, W.M.O. No. 438 (W.M.O., Geneva, 1976).
5. H. E. Landsberg, "Man-made climatic changes," *Science*, **170**, 1265–1274 (1970).
6. H. E. Landsberg, *Inadvertant Atmospheric Modification through Urbanization*, Technical Note No. BN 741 (Institute of Fluid Dynamics and Applied Mathematics, University of Maryland, 1972).
7. H. E. Landsberg, "Inadvertant atmospheric modification through urbanization," in *Weather and Climate Modification*, edited by W. N. Hess (Wiley, New York, 1974) pp. 726–763.
8. H. E. Landsberg, *Weather, Climate and Human Settlements*, W.M.O. Special Environmental Report No. 4 (W.M.O., Geneva, 1976).
9. M. E. Berlyand and K. Ya. Kondrat'yev, "Cities and the global climate," in *Meteorological Translations No. 23* (Atmos. Environ. Service, Canada, 1973) pp. 25–29.
10. W. R. Frisken, "The atmospheric environment of cities," in *The Atmospheric Environment* (Resources for the Future, 1973) pp. 9–48.
11. W. H. Terjung, "Urban climatology: with reference to the interrelationship between external weather and the microclimate in houses and buildings," in *Progress in Biometeorology*, Vol. 1, Part 1, edited by S. W. Tromp (Swets and Zeitlinger, Amsterdam, 1974) pp. 168–180.
12. T. R. Oke, "Inadvertant modification of the city atmosphere and the prospects for planned urban climates," in *Proceedings of a W.M.O. Symposium on Meteorology as Related to Urban and Regional Land-use Planning, Asheville*, W.M.O. No. 444 (W.M.O., Geneva, 1976) pp. 150–175.
13. T. R. Oke, "The climate of urban areas," in *Yearbook of Science and the Future* (Encyclopedia Britannica, Chicago, 1977) pp. 116–132.
14. W. P. Lowry, *North American Research Concerning the Effects of Urbanization on Local and Regional Weather and Climate*, Preprint of the Bilateral Seminar on the Geographical Character of Urban Development (Budapest, 1976).
15. R. R. Braham, Jr., *Overview of Urban Climate*, Paper presented to the S.U.N.Y./U.S.F.S./A.M.S. Conference on the Metropolitan Physical Environment (Syracuse, N.Y., 1975).
16. M. Daigo and T. Nagao, *Urban Climatology* (Asakura-Shoten, Tokyo, 1972).
17. K. Smith, *Principles of Applied Climatology* (McGraw-Hill, London, 1975) pp. 48–61.
18. M. M. Yoshino, *Climate in a Small Area* (University of Tokyo Press, 1975) pp. 15–16, 23–26, 80–119.
19. J. F. Griffiths, *Climate and the Environment* (Elek, London, 1976) pp. 125–131.
20. T. Kawamura, "Climatic modification by human activities," in *The Climate of Japan*, edited by E. Fukui (Kodansha–Elsevier, Tokyo–Amsterdam, 1977) pp. 209–224.
21. T. R. Oke, *Review of Urban Climatology, 1968–1973*, W.M.O. Technical Note No. 134, W.M.O. No. 383 (W.M.O., Geneva, 1974).
22. T. R. Oke, *The Urban Atmosphere as an Environment for Air Pollution Dispersion*, Preprint of a W.M.O. Meeting on Education and Training in Meteorological Aspects of Atmospheric Pollution and Related Environmental Problems (Research Triangle Park, North Carolina, 1977).
23. T. R. Oke, "The distinction between canopy and boundary layer urban heat islands," *Atmos.*, **14**, 268–277 (1976).
24. T. J. Lyons and B. W. Forgan, "Inadvertant weather modification by Chicago–North Indiana pollution sources observed by ERTS-1," *Mon. Weath. Rev.*, **102**, 503–508 (1974).
25. M. L. Wesley and R. C. Lipschutz, "An experimental study of the effects of aerosols on diffuse and direct solar radiation received during the summer near Chicago," *Atm. Env.*, **10**, 981–987 (1976).
26. F. Urbach, "Geographic pathology of skin cancer," in *The Biologic Effects of Ultraviolet Radiation* (Pergamon, Oxford, 1969) pp. 635–661.

27. D. Randerson, "A comparison of the spectral distribution of solar radiation in a polluted and a clear air mass," *J. Air Poll. Cont. Assoc.*, **20**, 546 (1970).
28. W. F. Dabberdt and P. A. Davis, "Determination of energetic characteristics of urban-rural surfaces in the Greater St Louis area," in *Preprints of a Symposium on Atmospheric Diffusion and Air Pollution, Santa Barbara* (American Meterological Society, Boston, 1974) pp. 133–141.
29. M. Ya. Berlyand, K. Ya. Kondrat'yev, O. B. Vasil'yev, I. V. Vasil'chenko, V. F. Zhalev, I. M. Zrazhevskii and L. S. Ivlev, "Combined investigations of peculiarities of the meteorological regime of a big city in the example of Zaporozh'ye (CENEX-72)," *Meteorol. Hydrol.*, No. 1, 14–23 (1974).
30. D. Yap, "Seasonal excess urban energy and the nocturnal heat island—Toronto," *Archiv. Met. Geophys. Biokl.*, Ser. B, **23**, 69–80 (1975).
31. T. R. Oke and R. F. Fuggle, "Comparison of urban-rural counter and net radiation at night," *Boundary-layer Met.*, **2**, 290–308 (1972).
32. W. R. Rouse, D. Noad and J. McCutcheon, "Radiation, temperature and atmospheric emissivities in a polluted urban atmosphere at Hamilton, Ontario," *J. Appl. Met.*, **12**, 198–807 (1973).
33. R. D. Bornstein, "Observations of the urban heat island effect in New York City," *J. Appl. Met.*, **7**, 575–582 (1968).
34. R. A. Ellefson, B. Takizawa and R. D. Bornstein, "Use of computer processed satellite data in numerical boundary layer models for land-use planning," in *Proceedings of W.M.O. Symposium on Meteorology as Related to Urban and Regional Land-use Planning, Asheville*, W.M.O. No. 444 (W.M.O., Geneva, 1976) pp. 261–279.
35. D. H. Yap and T. R. Oke, "Sensible heat fluxes over an urban area—Vancouver, B.C.," *J. Appl. Met.*, **13**, 880–890 (1974).
36. M. Nunez and T. R. Oke, "The energy balance of an urban canyon," *J. Appl. Met.*, **16**, 11–19 (1977).
37. K. Ya. Kondrat'yev and L. P. Orlenko (editors), *The Complex Atmospheric Energetics Experiment* (*Material of the CAENEX-70 Expedition*), Proceedings of the Main Geophysical Observatory, Voyeykov, No. 276 (1972).
38. K. Ya. Kondrat'yev, *The Complete Atmospheric Energetics Experiment*, GARP Public Series No. 12 (W.M.O., Int. Coun. Sci. Unions, Geneva, 1973).
39. K. Ya. Kondrat'yev, S. P. Malevsky-Malevich, L. R. Orlenko, N. E. Ter-Markaryanz, O. B. Vassilyev and V. F. Zhvalev, "Some results of investigations under the programme of the Complex Atmospheric Energetics Experiment (1970–1972)," *W.M.O.* Bulletin, **22**, 7–13 (1973).
40. L. B. Leopold, *Hydrology for Urban Land Planning: A Guidebook on the Hydrologic Effects of Urban Land Use*, U.S. Geological Survey Circular 554 (Washington, D.C., 1968).
41. H. W. Lull and W. E. Sopper, *Hydrologic Effects from Urbanization of Forested Watersheds in the Northeast*, U.S.D.A. Forest Service Research Paper NE-146 (1969).
42. W. L. Moore and C. W. Morgan, *Effects of Watershed Changes on Streamflow* (University of Texas Press, Austin, 1969).
43. J. C. Shaake, Jr., "Water and the city," in *Urbanization and Environment*, edited by T. R. Detwyler and M. G. Marcus (Duxbury, 1972) pp. 97–134.
44. Y.-T. Tam and R. D. Bornstein, *Anthropogenic Moisture Production and Its Effects on Boundary Layer Circulations over New York City*, Paper presented to S.U.N.Y./U.S.F.S./A.M.S. Conference on the Metropolitan Physical Environment (Syracuse, N.Y., 1975).
45. K. D. Hage, "Urban-rural humidity differences," *J. Appl. Met.*, **14**, 1277–1283 (1975).
46. D. L. Sisterson, *Studies on the Urban Moisture Budget*, Report No. AS114 (Department of Atmospheric Science, University of Wyoming, Laramie, 1975).
47. P. Halpern, C. Simon and L. Randall, "Source emissions and the vertically integrated mass flux of sulfur dioxide across the New York City area," *J. Appl. Met.*, **10**, 715–724 (1971).
48. F. L. Ludwig and J. H. S. Kealoha, *Urban Climatologic Studies*, Final Report, Contract COD-DAHC-20-67-C-0136 (Stanford Research Institute, California, 1968).

49. T. R. Oke, "City size and the urban heat island," *Atm. Env.*, **7**, 769–779 (1973).
50. J. R. Norwine, "Heat island properties of an enclosed multilevel suburban shopping center," in *Conference on the Urban Environment and Second Conference on Biometeorology* (American Meteorological Society, 1972) pp. 139–143.
51. F. A. Duckworth and J. S. Sandberg, "Vertical temperature differences observed over an urban area," *Bull. Amer. Met. Soc.*, **42**, 548–554 (1954).
52. J. F. Clarke, "Nocturnal urban boundary layer over Cincinnati, Ohio," *Mon. Weath. Rev.*, **97**, 582–589 (1969).
53. R. A. Dirks, "Urban atmosphere: Warm dry envelope over St Louis," *J. Geophys. Res.*, **79**, 3473–3475 (1974).
54. N. E. Bowne and J. T. Ball, "Observational comparison of rural and urban boundary layer turbulence," *J. Appl. Met.*, **9**, 862–873 (1970).
55. T. J. Chandler, *The Climate of London* (Hutchinson, London, 1965).
56. R. D. Bornstein and B. S. Johnson, "Urban–rural wind velocity differences," *Atm. Env.*, **11**, 597–604 (1977).
57. R. A. Baxter, Vertical Extent of Urban–Rural Velocity Differences, Unpublished manuscript (Department of Meteorology, San Jose State University, 1977).
58. W. P. Dannevik, S. Frisella, H. J. Hwang and R. B. Husar, *Analysis and Characterization of Heat-island-induced Urban Circulation during the 1973 E.P.A. Aerosol Characterization Study, St Louis, Missouri*, Preprint of the Fifth Conference on Weather Forecasting and Analysis (American Meteorological Society, Boston, 1974).
59. B. Ackerman, "Wind fields over the St Louis metropolitan area," *J. Air Poll. Cont. Assoc.*, **24**, 232–236 (1974).
60. T. Loose and R. D. Bornstein, "Observations of mesoscale effects on frontal movement through an urban area," *Mon. Weath. Rev.*, **105**, 563–571 (1977).
61. R. D. Bornstein, P. H. Fontana and W. T. Thompson, *Frictional Retardation of Sea Breeze Front Penetration in New York City and Its Implications for Pollutant Distribution*, Preprint of W. M. O. Symposium on Boundary Layer Physics Applied to Specific Problems of Air Pollution (Norköping, Sweden, 1978).
62. B. Ackerman, "Moisture content of city and country air," in *Preprints of a Conference on Air Pollution and Meteorology* (American Meteoreological Society, Boston, 1971) pp. 154–158.
63. R. D. Bornstein, A. Lorenzen and D. Johnson, "Recent observations of urban effects on winds and temperatures in and around New York City," in *Preprints of a Conference on the Urban Environment and the Second Conference on Biometeorology* (American Meteorological Society, Boston, 1972) pp. 28–33.
64. D. Sisterson and R. A. Dirks, *The Urban Moisture Climate*, Paper presented to S.U.N.Y./U.S.F.S./A.M.S. Conference on the Metropolitan Physical Environment (Syracuse, N.Y., 1975).
65. T. J. Chandler, "Absolute and relative humidities in towns," *Bull. Am. Met. Soc.*, **48**, 394–399 (1967).
66. H. E. Landsberg, "The climate of towns," in *Man's Role in Changing the Face of the Earth*, edited by W. L. Thomas (University of Chicago Press, 1956) pp. 584–606.
67. A. H. Auer, Jr., and R. A. Dirks, "Contributions to an urban meteorological study: METROMEX," *Bull. Am. Met. Soc.*, **55**, 106–110 (1974).
68. A. H. Auer, Jr., "The production of cloud physics Aitken nuclei by the St Louis metropolitan area," *J. Rech. Atm.*, **9**, 11–22 (1975).
69. R. R. Braham and P. Spyers-Duran, "Ice nucleus measurements in an urban atmosphere," *J. Appl. Met.*, **13**, 940–945 (1974).
70. A. J. Alkezweeny, "Evolution of the St Louis aerosol distribution," in *1974 Annual Report of the U.S. Atomic Energy Commission, Division of Biomedical and Environmental Research*, Part 3, Atmospheric Science Report BNWL-1950PT3 (Battelle Pacific Northwest Laboratories, Richland, Washington, 1975) pp. 65–69.
71. A. H. Auer, Jr., "Observations of an industrial cumulus," *J. Appl. Met.*, **15**, 406–413 (1976).

72. R. Cataneo, "Aircraft measurements and observations," in *Summary Report of METROMEX Studies, 1971–1972, Report of Investigation 74* (Illinois State Water Survey, Urbana, 1973) pp. 153–162.
73. W. Fitzgerald and P. A. Spyers-Duran, "Changes in cloud nucleus concentration and cloud droplet size distribution associated with pollution from St Louis," *J. Appl. Met.*, **12**, 511–516 (1973).
74. H. E. Dytch, Observations of Natural CCN and Large Aerosol over St Louis, in *Preprints of a Conference on Cloud Physics, Tucson* (American Meteorological Conference, Boston, 1974) pp. 508.
75. P. T. Schickedanz, "Inadvertant rain modification as indicated by surface raincells," *J. Appl. Met.*, **13**, 891–900 (1974).
76. H. T. Ochs, III, "Modeling of cumulus initiation in METROMEX," *J, Appl. Met.*, **14**, 873–882 (1975).
77. R. G. Semonin and S. A. Changnon, Jr., "METROMEX: Summary of 1971–1972 Results," *Bull. Am. Met. Soc.*, **55**, 95–100 (1974).
78. S. A. Changnon, Jr., "Inadvertant weather and precipitation modification by urbanization," *J. Irrig. Drain Div. A.S.C.E.*, **99 IRI**, 27–41 (1973).
79. D. F. Gatz, "METROMEX: Air and rain chemistry analyses," *Bull. Am. Met. Soc.*, **55**, 92–93 (1974).
80. R. P. Harnack and H. E. Lansberg, "Selected cases of convective precipitation caused by the metropolitan area of Washington, D.C.," *J. Appl. Met.*, **14**, 1050–1060 (1975).
81. B. W. Atkinson, *The Mechanical Effect of an Urban Area on Convective Precipitation*, Occasional Paper No. 3 (Department of Geography, Queen Mary College, University of London, 1975).
82. A. Detwiler, "Urban influence on cumulus formation," *Bull. Am. Met. Soc.*, **55**, 1240–1241 (1974).
83. W. A. Lyons, "Inadvertant weather modification by Chicago–Northern Indiana pollution sources observed by ERTS-1," *Mon. Weath. Rev.*, **102**, 503–508 (1974).
84. R. P. McNulty, "The effects of air pollutants on visibility in fog and haze at New York City," *Atm. Env.*, **2**, 625–629 (1968).
85. M. P. Paterson, "Visibility, humidity and smoke in Sydney," *Atm. Env.*, **7**, 281–290 (1973).
86. W. H. Terjung and S. S.-F. Louie, "Solar radiation and urban heat islands," *Ann. Assoc. Am. Geog.*, **63**, 181–207 (1973).
87. W. H. Terjung and S. S.-F. Louie, "A climatic model of urban energy budgets," *Georg. Anal.*, **6**, 341–367 (1974).
88. J. A. Arnfield, "Numerical modeling of urban surface radiative parameters," in *Papers in Climatology* (*Cam Allen Memorial Volume*), edited by J. A. Davies (McMaster University, Hamilton, 1976) pp. 1–28.
89. M. H. Unsworth, "Long-wave radiation at the ground. II. Geometry of interception by slopes, solids and obstructed planes," *Quart. J. Roy. Met. Soc.*, **101**, 25–34 (1975).
90. R. J. Cole, "The long-wave radiation incident upon the external surface of buildings," *Build. Serv. Eng.*, **44**, 195–206 (1976).
91. R. J. Cole, "The long-wave radiative environment around buildings," *Build. Env.*, **11**, 3–13 (1976).
92. M. Nunez, *The Energy Balance of an Urban Canyon*, Unpublished Ph.D. Thesis (University of British Columbia, Vancouver, 1975).
93. S. E. Nicholson, "A pollution model for street-level air," *Atm. Env.*, **9**, 19–31 (1975).
94. C. W. Newberry, K. J. Eatom and J. R. Mayne, "Wind loading on tall buildings—further results from Royex House," *Ind. Aerodyn. Abstr.*, **4**, 18 (1973).
95. N. Isyumov and A. G. Davenport, "Comparison of full-scale and wind tunnel wind speed measurements in the Commerce Court Plaza," *J. Ind. Aerodyn.*, **1**, 201–212 (1975).
96. A. D. Penwarden and A. F. E. Wise, *Wind Environment around Buildings*, Building Research Establishment Report (Department of the Environment, H.M.S.O., London, 1975).
97. P. A. Taylor, "Urban meteorological modeling—some relevant studies of turbulent diffusion in environmental pollution," in *Advances in Geophysics*, Vol. 18B, edited by F. N. Frenkiel and R. E. Munn (Academic, New York, 1974) pp. 173–185.

98. R. L. Lee, R. W. Bergstrom and R. D. Bornstein, "Workshop summary: Workshop on modeling the urban boundary layer, Las Vegas, Nevada, 5 May 1975," *Bull. Am. Met. Soc.*, **57**, 313–314 (1976).
99. C. C. Shir and R. D. Bornstein, "Eddy exchange coefficients in numerical models of the planetary boundary layer," *Boundary-layer Met.*, **11**, 171–185 (1977).
100. P. M. Tag, *Surface Temperatures in an Urban Environment*, Unpublished M.Sc. Thesis (McGill University, Montreal, 1964).
101. R. W. Bergstrom, Jr., and R. Viskanta, "Modeling of the effects of gaseous and particulate pollutants in the urban atmosphere. I. Thermal structure," *J. Appl. Met.*, **12**, 901–912 (1973).
102. R. W. Bergstrom, Jr., and R. Viskanta, "Modeling of the effects of gaseous and particulate pollutants in the urban atmosphere. II. Pollutant dispersion," *J. Appl. Met.*, **12**, 913–918 (1973).
103. R. W. Bergstrom, Jr., and R. Viskanta, "Prediction of the solar radiative flux and heating rates in a polluted atmosphere," *Tellus*, **25**, 486–498 (1973).
104. R. W. Bergstrom, Jr., and R. Viskanta, "Spherical harmonics approximation for radiative transfer in polluted atmospheres," *Prog. Astron. Aeron.*, **35**, 23–40 (1974).
105. A. Venkatram and R. Viskanta, *Radiative Effects of Pollutants on the Planetary Boundary Layer*, Environmental Monitoring Series, EPA-600/4-76-039 (U.S. Environmental Protection Agency, Research Triangle Park, North Carolina, 1976).
106. A. Venkatram and R. Viskanta, "Effect of elevated pollutant layers on mixed layer growth," in *Preprints of the Third Symposium on Atmospheric Turbulence Diffusion and Air Quality, Raleigh* (American Meterological Society, Boston, 1976) pp. 528–535.
107. A. Venkatram and R. Viskanta, "The contribution of pollutants to the urban heat island and 'crossover' effects," in *Preprints of the Third Symposium on Atmospheric Turbulence, Diffusion and Air Quality, Raleigh* (American Meteorological Society, Boston, 1976) pp. 536–542.
108. T. P. Ackerman, "A model of the effect of aerosols on urban climates with particular applications to the Los Angeles Basin," *J. Atm. Soc.*, **34**, 531–547 (1977).
109. T. P. Ackerman, K. N. Liou and C. B. Leovy, "Infrared radiative transfer in polluted atmospheres," *J. Appl. Met.*, **15**, 28–35 (1976).
110. J. Pandolfo and G. Wickham, *Modification of the TRC-USWB Boundary Layer Model*, Final Report 7474-180, Contract Cwb-10960 (The Travelers Research Center, 1965).
111. M. Lal, "Boundary layer model for thermal effects of pollutants in the atmosphere," *Archiv Met. Geophys. Biokl.*, Ser. B, **23**, 59–68 (1975).
112. M. Lal, "Radiative effects of pollutants in the atmospheric boundary layer," in *Preprints of the Third Symposium on Atmospheric Turbulence, Diffusion and Air Quality, Raleigh* (American Meteorological Society, Boston, 1976) pp. 543–546.
113. P. W. Summers, *An Urban Ventilation Model Applied to Montreal*, Unpublished Ph.D. Thesis (McGill University, Montreal, 1964).
114. F. Pasquill, "Prediction of diffusion over an urban area—current practice and future prospects," in *Proceedings of a Symposium on Multisource Urban Diffusion Models* (U.S. Environmental Protection Agency, Research Triangle Park, North Carolina, 1970) pp. 3.1–3.25.
115. D. M. Leahey, *An Urban Heat Island Model*, Final Report TR-69-11 (New York University, 1969).
116. D. M. Leahey and J. P. Friend, "A model for predicting depth of the mixing layer over an urban heat island with applications to New York City," *J. Appl. Met.*, **10**, 1162–1173 (1971).
117. J. D. Kalma, "An advective boundary-layer model applied to Sydney, Australia," *Boundary-layer Met.*, **6**, 351 (1974).
118. E. Gold, "Smog. The rate of influx of surrounding cleaner air," *Weather*, **11**, 230–232 (1956).
119. B. F. Findlay and M. S. Hirt, "An urban-induced mesocirculation," *Atm. Env.*, **3**, 537–542 (1969).

120. N. E. Bowne and J. T. Ball, "Observational comparison of rural and urban boundary layer turbulence," *J. Appl. Met.*, **9**, 862–873 (1970).
121. F. M. Vukovich, "Theoretical analysis of the effect of mean wind and stability on a heat island circulation characteristic of an urban complex," *Mon. Weath. Rev.*, **99**, 919–926 (1971).
122. D. B. Olfe and R. L. Lee, "Linearized calculations of urban heat island convection effects," *J. Atm. Sci.*, **28**, 1374–1388 (1971).
123. T. Yamada and R. N. Merony, *Numerical and Wind Tunnel Simulation of Response of Stratified Shear Layers to Nonhomogeneous Surface Features*, Project THEMIS Report No. 9 (Colorado State University, Fort Collins, Colorado, 1971).
124. R. N. Meroney and T. Yamada, "Wind tunnel and numerical experiments of two-dimensional stratified airflow over a heated island," in *Proceedings of A.M.S. 92nd Winter Annual Meeting* (1971) pp. 31–40.
125. R. N. Meroney and T. Yamada, "Numerical and physical simulation of a stratified airflow over a series of heated islands," in *Proceedings of Summer Computer Simulation Conference*, Vol. 2 (A.I.A.A./A.I.Ch.E./A.M.S./I.S.A./S.C.I./S.H.A.R.E., 1972) pp. 910–917.
126. T. Yamada, "Urban heat island effects on air pollution," in *Preprints of a Conference on the Urban Environment and the Second Conference on Biometeorology* (American Meteorological Society, Boston, 1972) pp. 83–88.
127. M. A. Estoque, "A numerical model of the atmospheric boundary layer," *J. Geophys. Res.*, **68**, 1103–1114 (1963).
128. J. L. McElroy, *An Experimental and Numerical Investigation of the Nocturnal Heat Island over Columbus, Ohio*, Unpublished Ph.D. Thesis (The Pennsylvania State University, 1971).
129. J. L. McElroy, "A numerical model of the nocturnal urban boundary layer," in *Proceedings of a Symposium on Air Pollution, Turbulence and Diffusion* (Las Cruces, New Mexico, 1972).
130. J. L. McElroy, "Effects of alternate land use strategies on the structure of the nocturnal urban boundary layer," in *Preprints of a Conference on the Urban Environment and the Second Conference on Biometeorology* (American Meteorological Society, Boston, 1972) pp. 185–190.
131. J. L. McElroy, "A numerical study of the nocturnal heat island over a medium-sized midlatitude city (Columbus, Ohio)," *Boundary-layer Met.*, **3**, 442–453 (1973).
132. N. K. Wagner and T. Yu, "Heat island formation: a numerical experiment," in *Preprints of a Conference on the Urban Environment and the Second Conference on Biometeorology* (American Meteorological Society, Boston, 1972) pp. 83–88.
133. T.-W. Yu, *Two-dimensional Time-dependent Numerical Simulation of Atmospheric Flow over an Urban Area*, Report No. 42 (Atmospheric Science Group, University of Texas, 1973).
134. R. D. Bornstein, *Two-dimensional Nonsteady Numerical Simulations of Night-time Flows of a Stable Planetary Boundary Layer over a Rough Warm City*, Ph.D. Thesis (Department of Meteorology and Oceanography, New York University, 1972).
135. R. D. Bornstein, "Two-dimensional nonsteady numerical simulation of night-time flow of a stable planetary boundary layer over a rough warm city," in *Preprints of a Conference on the Urban Environment and the Second Conference on Biometeorology* (American Meteorological Society, Boston, 1972) pp. 89–94.
136. R. D. Bornstein, "The two-dimensional URBMET urban boundary layer model," *J. Appl. Met.*, **14**, 1459–1477 (1975).
137. T. R. Oke and C. East, "The urban boundary layer in Montreal," *Boundary-layer Met.*, **1**, 411–437 (1971).
138. J. Pandolfo, D. Cooley and M. Atwater, *The Development of a Numerical Prediction Model for the Planetary Boundary Layer*, Final Report 7465-174, Contract Cwb-10960 (The Travelers Research Center, 1965).
139. J. K. Angell, D. H. Pack, C. R. Dickson and W. H. Hoecker, "Urban influence on night-time air flow estimated from tetroon flights," *J. Appl. Met.*, **10**, 194–204 (1971).

140. D. Dieterle, *Simulation of the Urban Surface Energy Balance, Including the Effects of Anthropogenic Heat Production*, Unpublished M.Sc. Thesis (San Jose State University, California, 1979).

141. D. P. Gutman, *Heat Rejection and Roughness Effects on the Planetary Boundary Layer above Cities*, Unpublished Ph.D. Thesis (Cornell University, Ithaca, N.Y., 1974).

142. D. P. Gutman and K. E. Torrance, "Response of the urban boundary layer to heat addition and surface roughness," *Boundary-layer Met.*, **9**, 217–233 (1975).

143. R. L. Lee and D. B. Olfe, "Numerical calculations of temperature profiles over an urban heat island," *Boundary-layer Met.*, **7**, 39–52 (1974).

144. D. C. Barnum and G. V. Rao, "The interaction of the urban heat island with the planetary boundary layer," in *Preprints of the Fifth Conference on Weather Forecasting and Analysis* (American Meteorological Society, Boston, 1974) pp. 176–179.

145. D. C. Barnum and G. V. Rao, "Role of advection and penetrative convection in affecting the mixing-height variations over an idealized metropolitan area," *Boundary-layer Met.*, **8**, 497–514 (1975).

146. J. F. Black *et al.*, *A Nonlinear Three-dimensional Steady-state Model of Convection over a Heat Island in the Presence of an Imposed Wind* (Esso Research and Engineering, Linden, N.J., 1971).

147. J. P. Pandolfo *et al.*, *Predictions of Numerical Models of Transport and Diffusion in an Urban Boundary Layer*, Final Report, Vol. 1, Contract No. 4082 (The Center for the Environment and Man, Hartford, Conn., 1971).

148. M. A. Atwater, "The radiation budget for polluted layers of the urban environment," *J. Appl. Met.*, **19**, 205–214 (1971).

149. M. A. Atwater, "Radiative effects of pollutants in the atmospheric boundary layer," *J. Atm. Sci.*, **28**, 1367–1373 (1971).

150. M. A. Atwater, "Thermal effects of urbanization and industrialization in the boundary layer. A numerical study," *Boundary-layer Met.*, **3**, 229–245 (1972).

151. M. A. Atwater, "Thermal changes induced by urbanization and pollutants," *J. Appl. Met.*, **14**, 1061–1071 (1975).

152. M. A. Atwater and J. P. Pandolfo, "Tundra environmental changes induced by urbanization," in *Climate of the Arctic* (Geophysical Institute, University of Alaska, Fairbanks, Alaska, 1975) pp. 312–315.

153. F. M. Vukovich, J. W. Dunn, III, and B. W. Crissman, "A theoretical study of the St Louis heat island: The wind and temperature distribution," *J. Appl. Met.*, **15**, 417–440 (1976).

154. M. L. Davis, *Modeling Urban Atmospheric Temperature Profiles*, Unpublished Ph.D. Thesis (University of Illinois, Urbana, 1968).

155. M. L. Davis and J. E. Pearson, "Modeling urban atmosphere temperature profiles," *Atm. Env.*, **4**, 277–288 (1970).

156. N. J. Cook, "On simulating the lower third of the urban adiabatic boundary layer in a wind tunnel," *Atm. Env.*, **7**, 691–705 (1973).

157. N. J. Cook, "On applying a general atmospheric simulation method to a particular urban site for *ad hoc* wind loading or wind environment studies," *Atm. Env.*, **8**, 85–87 (1974).

158. J. Counihan, "Simulation of an adiabatic urban boundary layer in a wind tunnel," *Atm. Env.*, **7**, 673–689 (1973).

159. S. SethuRaman, *Stratified Shear Flows over a Simulated Three-dimensional Urban Heat Island*, Unpublished Ph.D. Thesis (Colorado State University, Fort Collins, 1973).

160. S. SethuRaman, "Air mass modification due to change in surface characteristics," *Mon. Weath. Rev.*, **104**, 1040–1043 (1976).

161. S. SethuRaman and J. E. Cermak, "Physical modeling of flow and diffusion and pollution," in *Advances in Geophysics*, Vol. 18B, edited by F. N. Frenkiel and R. E. Munn (Academic, New York, 1974) pp. 223–240.

172. S. SethuRaman and J. E. Cernak, "Mean temperature distributions over a physically modeled three-dimensional heat island for different stability conditions," in *Preprints of a Symposium on Atmospheric Diffusion and Air Pollution, Santa Barbara* (American Meteorological Society, Boston, 1974) pp. 381–386.

163. S. SethuRaman and J. E. Cermak, "Mean temperature and mean concentration distributions over a physically modeled three-dimensional heat island for different stability conditions," *Boundary-layer Met.*, **9**, 427–440 (1975).

Robert D. Bornstein
San Jose State University
and Tim Oke
The University of British Columbia

ELECTROSTATIC PRECIPITATORS FOR PARTICULATE CONTROL

Electrostatic precipitators have been used to remove and recover fine solids and liquids from gas streams for the better part of this century. The definitive work by White, *Industrial Electrostatic Precipitation*, presents a far better and more detailed discussion of the history of precipitators and their application than space permits here.[1] Those who wish more detailed presentation should consult the authoritative works,[2–4] and also the more recent discussions which bring these up to date and provide further insights into the theory and practice of high-resistivity precipitation on the basis of experimental data gathered in the last decade.[5,6]

Until the 1960s, electrostatic precipitators—henceforth, for brevity, precipitators—were justified in terms of public goodwill, the elimination of unsightly smoke or the recovery of valuable material such as metal fines. With the advent of the environmental era, precipitators became the device for particulate control. At first, units operated at removal efficiencies of between 90 and 95%. As the requirements to meet the Ambient Air Quality Standards increased and discharge limits were tightened, precipitators were designed for and operated at efficiencies in excess of 99%. Today, precipitators are designed to remove in excess of 99.5% of the particles presented to the unit. The units, when adequately sized, have no trouble in meeting utility-emission standards of 0.1 lb, particulate per million Btu of fuel fired.

A number of recent trends affect the design and economics of precipitators, most notably the increasingly stringent particulate-emission standards and the desire to burn low-sulfur fuels in order to meet SO_x-emission standards. The latter, along with the general changeover to coal from oil, has led to coal switching. Generally the coals produce ash of higher resistivity and are thus more difficult to collect. This necessitates the use of a larger precipitator to obtain the desired efficiency, or process modifications to improve the resistivity and collectability of the ash. The decrease in permitted emissions also necessitates a larger precipitator or similar process modifications. In some selected cases where emission requirements are extremely tight, particularly with respect to submicron particles, a traditional precipitator may no longer be economical. In these cases, either a fabric filter or a precipitator using ash modification by chemical conditioning may be more suitable. Each of these marginal cases must be examined on its own merits.

As the standards become more stringent and coal switching more prevalent, the major challenges to continued utilization of the precipitator are solutions to the dust-reentrainment and high-resistivity problems. These are discussed in detail below.

Rapping and Dust Collection

In most precipitators, rapping is utilized to remove dust from the collecting surfaces. In limited industrial applications it is possible to use wetted wall precipitators, which do not require rapping; these will not be discussed here. The intensity and mode of rapping may be key factors in attaining high efficiencies, because these factors affect the way in which dust is moved from the plate. At one time it was thought that out-of-plane accelerations greater than 50*g* were required to move the dust along the plate into the hopper when high-resistivity ashes were handled. Early work showed that, the more intense the rapping, the easier the dust moved.[7,8] But it was also shown that high-intensity rapping could increase reentrainment.[7]

Recent experimental work indicates that optimal dust removal is a function of in-plane as well as out-of-plane accelerations.[9,10] Other work has shown that the rapping response is extremely complex; it lasts for several milliseconds at any point and consists of a series of peaks and harmonics.[11] To date, analysis of the quality of the plate response and its effect on dust removal has not been possible.

A collecting surface has a wide *g* response. The maximum *g* accelerations at any given point vary immensely as the same power of the distance from the rap-impact point. A typical map of acceleration lines is shown in Figure 1. Complex as this is, it does not provide an insight into the effect of dust properties discussed in detail by Tassicker.[12]

The nature of the plate response when a rapping blow is struck is extremely complex. Any portion of the collecting surface reacts to the blow in ways that are determined by its own physical properties and those of its surrounding bounding elements. Flat plates theoretically have an infinite number of modes of vibration. Constraints of shape, such as bends, baffles or joints, contribute in defining the frequency content of any portion of the collecting surface, just as shaping oil-drum ends produces the musical instrument featured by the Caribbean steel bands.

The collecting surface is like a huge steel drum. The tuning of the frequency response is determined indirectly by the choice of features on the basis of

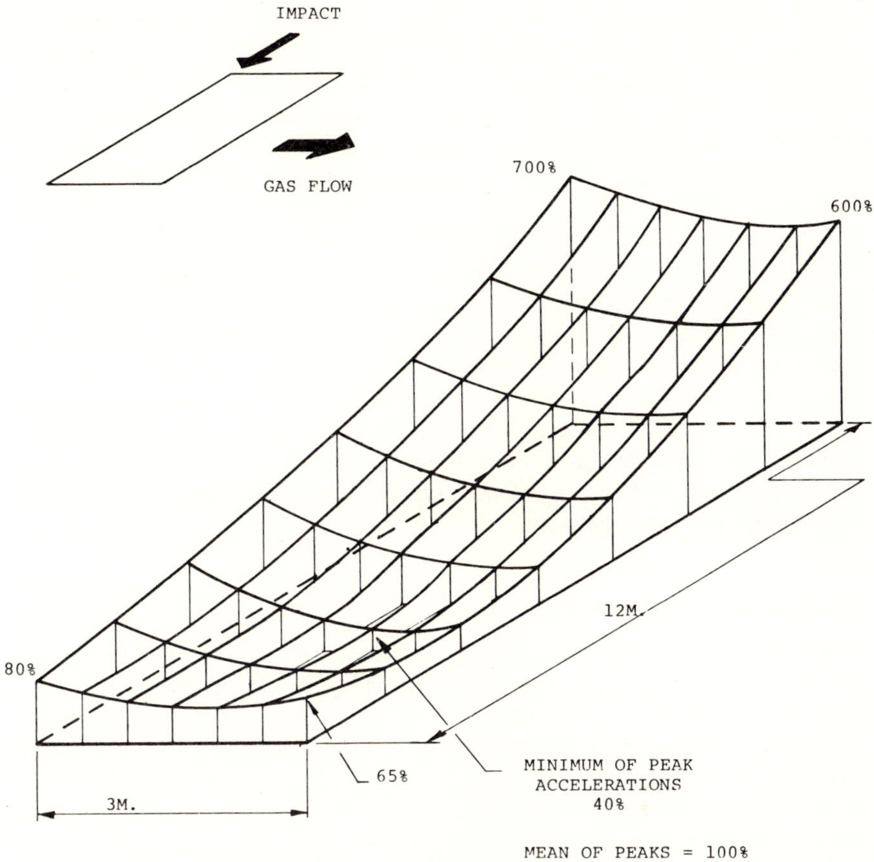

Figure 1 Topographical map of typical peak out-of-plane accelerations.

particulate-collection efficiency and manufacturing practicality. And when it is hit, it rings. Although a rapping blow with a steel hammer lasts less than a thousandth of a second—the time of contact between the hammer or rapper weight and the struck system—it is evident from oscilloscope traces showing the response of an accelerometer fixed to a point on a collecting surface that out-of-plane surface motion continues for as long as 200 ms after the response to the blow begins.

Thus, the response to a rapping blow at any section of the collecting surface is not uniform. The action of a dropped 20-lb hammer can produce a force pulse of about 14,000 lb, which, when acting on 2000 lb of collecting surface, could only develop a 7*g* acceleration peak. The accelerations reported in the precipitator result from the local vibration responses.

The acceleration–time trace, processed by a real-time frequency-spectrum analyzer, shows that the principal contributors to the acceleration amplitudes derive from the frequency range 1500–2000 Hz. Out-of-plane accelerations of more than 1000g are reported. This is not realistic, as is apparent on recognizing that the accelerations corresponding to displacements at these frequencies are numerically $(2\pi f)^2/g$ as large as the displacement itself. If the plate should have a single response frequency of 1750 Hz, the 200 ms of vibration activity would include some 700 positive or negative acceleration interactions of varying amplitude between the collecting surface and the layer of collected dust. There is a similar scenario for in-plane accelerations. And very clearly, the dust and plate interface is stressed to breaking by both the in-plane and out-of-plane accelerations. Which type of accelerations and which peaks are the critical ones has yet to be established. Further difficult experimental studies are required before the true nature of the plate–dust interface and interactions are understood. Only then can the various plate designs and rapping modes on the market be truly compared. In the meantime, it is important not to oversimplify.

Rapping and Reentrainment

Typically, large modern utility precipitators operate at a removal efficiency of 99.5%. For most operations such efficiency would be remarkable; for precipitators greater efficiencies must be attained to meet new, stringent, air-quality regulations. Every particle that is not collected or is reentrained into the gas stream is critical.

Very little meaningful quantitative information is available on the factors that affect reentrainment. This is because there are very many uncontrolled variables in the operation of a power-boiler precipitator that affect particle reentrainment. Among the more critical variables are:

1) the dust resistivity;
2) the gas velocity;
3) the particle cohesiveness;
4) the rapping frequency and intensity;
5) the corona power levels;
6) the hopper flow and dust backup.

A study by Spencer *et al.* showed that the losses from a high-efficiency precipitator were largely due to rapping reentrainment,[13] but neither this nor any other has presented data which might provide a basis for the design of a plate-rapping system to minimize reentrainment.

Some plate experiences shows that, in cases where dust resistivity is very favorable, reentrainment is minimal and does not appear to depend heavily on the frequency and intensity of plate rapping. When the dust resistivity is unfavorably low, the collected particles can be easily reentrained at relatively low plate-rapping energies.

Some indications exist that there is an optimum frequency and intensity for plate rapping which depends on the concentration of ash in the gas stream as well as the variables previously discussed. If rapping is too frequent, the particles will have insufficient time to agglomerate on the plate and so will be dislodged back into the gas stream. If rapping is too infrequent, a large cake of ash will form on the plate. When the plate is rapped, there is an avalanche of ash, which separates from the plate, and thus a large portion of the dislodged ash reentrains. According to one study, an optimum rapping frequency was achieved by allowing 1/8 in of ash to accumulate on the plates between raps.

It has been found that high precipitator-treatment velocity is one of the most significant contributors to dust reentrainment, particularly for low-resistivity applications.

Increasing the treatment velocity from 4.5 to 6.0 ft/s was found to increase emission by a factor of 10 in one case. At 6.0 ft/s, 50% of the emission was attributable to rapping, while most of the remaining particulate was reentrained from the plates between raps as a result of the high treatment velocity. High-resistivity particulate is usually less sensitive to the treatment velocity, since the electrostatic forces holding the particle to the plate are usually greater if back corona is not present. However, higher treatment velocities will increase the probability that a particle reentrained through rapping will escape the precipitator.

High-resistivity Dust

One of the most serious problems facing modern electrostatic precipitation is the collection of high-resistivity dust, particularly the fly ash resulting from the combustion of low-sulfur coal. The increasingly stringent emission standards for fly ash compound the difficulties, because the very high particulate-collection efficiencies required to meet new standards, in combination with the high-resistivity problem, could necessitate the use of very large and expensive precipitators. Such applications require a breakthrough in precipitation technology so as to be able to overcome the resistivity problem and to meet the high efficiency requirements with precipitators of reasonable size. Several approaches to this problem are presently being pursued. These will be described later, following a brief discussion of electrostatic-precipitation theory necessary to understand the resistivity problem.

Theory

Electrostatic precipitators remove particulate emissions from gas streams by first charging the particles by corona discharge and then using the electrical force produced by the electric field to cause the particles to migrate to a collecting surface. Industrial electrostatic precipitators are generally of single-stage design, meaning that a single-electrode geometry is used throughout the precipitator to accomplish both charging and collecting functions. The electrode geometry normally used consists of a series of discharge wires located between parallel collecting plates. The plates are grounded and the wires are at high negative potential. Typically, the wires may be 2–3 mm in diameter and spaced 14 cm apart. The collecting-plate spacing is 20–30 cm. Voltage levels as high as 50 kV on average can be applied in this geometry, although operation is normally limited to lower levels for the reasons discussed below.

As high voltage is applied to the discharge wires, corona discharge occurs when the applied voltage exceeds the corona-initiation voltage, normally between 20 and 30 kV in the geometry described above. This discharge results in the formation of negative ions, which then migrate toward the collecting plate (anode) under the influence of the electric field existing between the wire and the plate.

Particles flowing through this electric field tend to distort the field somewhat. Ions are then attracted to the particle along the resultant lines of force, thus imparting their charge to the particle. As the particle takes on more and more charge, the lines of force eventually bypass the particle completely when the particle has reached its saturation charge. This maximum charge is proportional to the applied electric field strength and to the square of the particle diameter.

The rate of particle charging depends on the density of ions surrounding the particle. Normally, this rate is very rapid, a particle attaining over 90% of the saturation charge in a fraction of a second.

The process described above is known as "field charging." It is effective primarily for particles larger than a few microns in diameter. For submicron particles, charging by diffusion of ions to the particles becomes the dominant mechanism. As with field charging, the rate of charging depends on the ion density, but in this, case unlike field charging, the charge is not subject to an upper limit which depends on the field strength. In general, more time is needed for effective charging by diffusion than for field charging.

Charged-particle motion toward the collecting plates in the precipitator is influenced mainly by (1) the electrical force on the particles; (2) viscous drag on the particle moving relative to the gas stream; and (3) turbulence of the gas stream. The electrical force on a particle is caused by the electric field in the vicinity of the particle. There are two components to this electric field strength, one due to the voltage difference between the anode and the cathode and their relative geometries, and the other due to the space charge created by ions and other charged particles in the interelectrode space. The force on the particle is in the direction of the collecting plates and is proportional to the charge on the particle and the electric field strength. The motion toward the collecting plate caused by the electrical force is opposed by viscous drag on the particle. By force balance, an equation for particle velocity toward the plate can be derived:

$$w_d = \frac{QEC}{3\pi\mu d},$$

where w_d is the particle-migration velocity, m/s; Q is the particle charge, C; E is the electric field strength, V/m; C is the Cunningham correction factor; μ is the gas viscosity, dpoise; d is the particle diameter, m.

Flow in industrial precipitators is very turbulent. This turbulence plays a strong role in determining the particle motion, and it must be considered in any modeling of the precipitation process in order to estimate the rate of particle collection. Many models have been proposed for the prediction of particulate collection in an electrostatic precipitator. They differ among themselves primarily in the way the turbulent diffusion of the particles is treated and in the boundary conditions assumed. Although special advantages can be argued for each, perhaps the most useful model is the Deutsch–Anderson model, because of its applicability and ease of use. Essentially, the Deutsch–Anderson model assumes complete lateral mixing because of high turbulence in the gas stream between the collecting plates. This means that there is no particulate-concentration gradient transverse to the flow at any point in the precipitator; the turbulence destroys any gradient which would tend to develop because of net particle migration toward the plates. Therefore, electrical migration of the particles is important only across the boundary layer at the collecting plate. The resultant equation for the precipitator collection efficiency using this model is:

$$\eta_d = 1 - e^{-w_d L/vS},$$

where η_d is the fraction of particles collected that have diameter d; L is the precipitator length, m; v is the gas velocity, m/s; S is half the plate spacing, m.

It is important to note the dependence of the collection efficiency on particle size because of the dependence of w_d on the particle size. In actual applications; however, the precipitator must deal with particulate emissions covering a wide range of particle sizes; for example, the size distribution of particulate emissions from a pulverized-coal fixed boiler is characterized by a mean diameter of 12 μm with a geometric standard deviation of 3.8. In spite of significant variations in individual w_d over the size range of particulate emissions, the Deutsch–Anderson equation is used extensively to predict and correlate precipitator performance in various applications using a single value of w known as the "effective migration velocity." Unfortunately, this quantity changes from one efficiency level to another in any given application, because the particle size distribution changes as more particles are collected. One way to avoid this difficulty in dealing with log-normal particle size distributions is to use an efficiency equation of the form:

$$\eta = 1 - c^{(w'L/vS)^m}.$$

Here c and m are constants dependent only on the *inlet* size distribution and w' is a migration velocity independent of particle size. This equation was derived by integrating the Deutsch–Anderson equation over the particle size distribution using the individual w_d for each particle size. The derivation and advantages of this form of the Deutsch–Anderson equation are given by Feldman.[14] It can be shown that, typically,

$$w' = (2.95 \times 10^{-16}) E_0 E_p/\mu,$$

where E_0 is the particle-charging field strength and E_p is the field strength for particle collection. In the Deutsch–Anderson model, E_p is the field at the precipitator collecting surface, while E_0 is an average field strength in the interelectrode space. Thus, w' is a direct measure of the electrical energization of the precipitator independent of particle-size considerations. For this reason, it is a useful parameter when studying the effects of high-resistivity dust in limiting the electrical energization of a precipitator.

The Problems of High Resistivity

The electrical resistivity of the particulate collected in an electrostatic precipitator is very important in determining the performance of the precipitator. In various applications, it is possible to encounter resistivities ranging from less than 10^7 Ω-cm to greater than 10^{13} Ω-cm. The range greater than about 2×10^{10} Ω-cm is termed high resistivity. It is in this range that the electrical energization of a precipitator is adversely affected by the resistivity of the dust layer on the collecting plate; the problem becomes worse as resistivity increases.

Typically, resistivity–temperature curves for fly ash show a maximum in resistivity at about 150°C; this is typical of flue-gas temperatures to the precipitator. The reason for the resistivity maximum is that there are two primary mechanisms for electrical conduction in a dust layer, one favoring lower temperatures and the other favoring high temperatures. They are volume conduction and surface conduction.

Volume conduction refers to the conduction of charge through the individual particles. For most materials, volume conduction increases with increasing temperature; this accounts for the decreasing resistivity curve on the high-temperature side of the maximum.

Surface conduction refers to the conduction of charge through an adsorbed or condensed film on the surface of the particles. The adsorption and condensation of moisture increases as the temperature decreases, resulting in the decreasing resistivity curve on the low-temperature side of the maximum. Some chemical compounds in the flue gas also influence the adsorption or condensation process and therefore affect the resistivity curve. For example, the addition of traces of sulfur trioxide to the flue gas causes a significant reduction in resistivity as a result of the formation of a sulfuric-acid film on the particulate surface.

When a precipitator collects high-resistivity dust, the main symptom of the problem is poor electrical energization of the precipitator, manifested either as low power input as a result of excessive sparking at low voltages or high power input as a result of back corona. The reason for this behavior is that, when a poorly conductive ash layer develops on the precipitator plates, it is characterized by a significant voltage gradient, equal to the product of the resistivity and the current density through the layer. For fly ash, a voltage gradient in the ash layer of more than about 10 kv/cm leads to dielectric breakdown of the layer. For very high resistivity levels (about 10^{12} Ω-cm or more), dielectric breakdown occurs very close to the start of the normal corona. In this case, the field strength in the gap is too low to cause a spark to propagate across the gap and the breakdown appears as stable back corona, i.e., the generation of positive ions in the ash layer and consequent typical rapid rise in current density. Precipitation suffers for several reasons when back corona occurs. The low voltage at which back corona starts severely limits the electric field strength, which is important in both particle charging and particle collection. The presence of positive ions as a result of the reverse ionization undermines the particle-charging process and further tends to reduce the component of electric field strength normally due to negative space charge. In addition there is evidence that, as back-corona current increases, the average static field strength tends to decrease, further compounding the problem.

When resistivity is only moderately high (about 10^{11} Ω-cm), electrical energization still suffers. However, higher voltages and useful currents, as compared to the case of very high resistivity, are achieved before dielectric breakdown occurs. In this case, the breakdown appears as sparking rather than back corona because the somewhat higher field strengths allow streamers to propagate across the gap. The precipitation is still far from optimum, because the spark limitation is reached at low current densities and relatively low voltages. Thus, the particle-charging and particle-collecting fields and ion densities are not high enough for good precipitation. Another adverse factor accompaning low-level energization, whose importance is often overlooked, is the corona distribution on the discharge electrode: with high resistivity, the corona points are very sparse and poorly distributed along the length of each electrode, further reducing the average ion density seen by the particles as they are charged.

The effect of high-resistivity operation on w' can be quite severe. For example, electric field strengths in high-resistivity operation may in some cases be only 70% of those attainable in lower-resistivity operation. As seen before, this affects both particle charging and particle collection, the net effect on w' being proportional to the square of the field-strength reduction. Thus, in this example, w' would be only about 50% of the value for well-energized operation. This means that a precipitator designed for this high-resistivity case would have to be twice as large as one designed for operation below the high-resistivity range for any given collection efficiency. The impact on precipitator economics and acceptability is severe, especially for the very high efficiencies required today. Solutions to the problem are necessary if precipitation is to remain a viable alternative for particulate control in high-resistivity applications.

Solutions to the High-resistivity Problem

There are two general methods of effectively avoiding the problems of high resistivity in electrostatic-precipitator operation. The most obvious is to alter the resistivity of the dust layer on the anode. Examples of this approach will be discussed below; they include chemical conditioning, hot precipitation and wet precipitation. The second general method is to energize the precipitation in such a way that limitations due to the high-resistivity dust layer are minimized. An example of this approach is pulsed energization. Most of these approaches can be used effectively and economically to avoid the poor performance of electrostatic precipitators because of high resistivity.

Hot Precipitators

The major application of electrostatic precipitation is in collecting fly ash from coal-fired utility and industrial boilers. Normally, the precipitator is located just downstream of the air preheater, where flue-gas temperature is typically 120–200°C. In this temperature range, resistivity is typically at its highest. Examination of the resistivity curve suggests that the resistivity can be lowered by operating in either a lower or higher temperature range. Operation at lower temperatures is not generally practical. However, higher-temperature operation can be achieved simply by placing the precipitator ahead of the air preheater, where temperatures range from 300 to over 400°C. In this range, fly ash resistivity is normally not a problem. Precipitators operating in this temperature range are commonly termed *hot precipitators.*

Hot precipitators are widely used by the utilities, and they do accomplish the objective of allowing electrostatic precipitation outside of the high-resistivity range. Where applicable, hot precipitators are an economical choice over conventional precipitators. However, there are some disadvantages to the use of hot precipitators which must be weighed in making a choice of control alternatives. Because of the higher temperature, the actual gas volume handled by the precipitator is larger by as much as 70%. Much of the precipitator-size reduction gained by avoiding high-resistivity operation can be lost as a result of the higher volume.

Construction materials for hot precipitators are generally more expensive than for conventional precipitators. Power supplies must be designed to provide greater current densities than conventional operation because of the effect of lower gas density at the higher temperature.

Another disadvantage of hot precipitation is that heat loss from the precipitator may reduce the temperature to the air preheater, thus affecting the overall thermal efficiency of the power-generation process. In spite of these disadvantages, hot precipitation is often a practical, cost-effective solution to the high-resistivity problem.

Wet Precipitation

Another means of operating the precipitator outside the high-resistivity range is to introduce water and operate with wetted plates. Because of the presence of liquid water, the high resistivity of the dust will not be a problem. In addition, dust reentrainment will also be minimized if the plates are wet. Despite this double advantage, wet precipitation is not widely employed because of its accompanying problems.

The operation of a completely wet precipitator requires saturation of the flue gas. In most applications, especially with utility boilers, this would require a large supply of water, and would have the additional adverse effect of causing a loss of buoyancy of the gas to the stack; additional power would be needed to reheat the stack gas. Further problems include the need for special construction materials for wet operation and the disposal of the wet sludge from the precipitator.

At present, the problems of wet precipitation outweigh the advantages for many applications, including utility and industrial boilers. However, it is quite likely that wet precipitation will eventually find use as part of a total system for the combined removal of particulates and gaseous pollutants.

Pulsed Energization

This method of dealing with high-resistivity dust involves a change in the way the precipitator is energized. The actual resistivity of the dust layer is not affected, although the ensuing limitations are minimized. Pulsed energization has been recognized as a desirable advance in the art of electrostatic precipitation for a very long time; patents on the subject were issued as long ago as the 1930s. However, commercialization of the concept has been slow, partly because of certain technical and economic problems, but also because the need for highly efficient precipitation of high-resistivity dust has not been felt until recently.

Several systems for pulsed energization are known to be presently under development. A system being developed by Research Cottrell and the Ion Physics Co. superimposes high-voltage impulses of very short duration and steep wavefront on an underlying, relatively constant potential. This constant "base voltage" is maintained at a level just below the onset of sparking or back corona and provides a field to maintain ion and particle migration toward the collecting plates. In the case of very high resistivity, this base voltage will preferably be below the normal corona starting voltage. The high-voltage impulses momentarily raise the total potential well above the sparking or back corona limit. This pulse voltage can actually be 2–3 times the base voltage because breakdown of a gas is high with respect to short pulsed potentials. The effect of the high-voltage impulses is to initiate strong and uniform corona generation but, by properly adjusting important pulse parameters such as the pulse width and frequency, the adverse effects of excessive sparking and back corona can be avoided.

The net result is a significant improvement in the current density and in the uniformity of the current over conventional energization. The precipitation process benefits from the higher and more uniform current density through a greater availability of ions for particle charging. Furthermore, the electric field strengths necessary for particle charging and collection are improved. Both the collection and charging fields are improved by the increased space charge, while the charging field enjoys an additional improvement as a result of the instantaneously high field strength present during the pulse.

The preliminary economics of this system show it to be a very attractive means of dealing with high resistivity. Its eventual commercialization may prove significant in maintaining electrostatic precipitation as a viable alternative for the collection of high-resistivity dust.

Two-stage Precipitator

Conventional industrial precipitators are of the single-stage design, i.e., both particle charging and particle collection are accomplished by the same electrode geometries. In a two-stage precipitator, the particles are first charged in a section specifically designed for optimum charging, and are then collected in downstream sections specifically designed for optimum collection. This two-stage concept is particularly applicable to precipitators dealing with high-resistivity dust, for reasons discussed below.

Since particle charging is a very rapid process, the volume of a precipitator which must be devoted to charging can be a small fraction of the total precipitator. For effective charging, both high ion density and high electric field strength are desired, but high resistivity limits the extent to which these conditions can be realized. As discussed above, heating, cooling or wetting of the dust layer can counteract the resistivity limitation but, except for the hot precipitator, none of these approaches is generally practical in the single-stage precipitator. However, for particle charging only, these approaches may indeed be practical, especially if the anode design is such as to minimize the total anode area where dielectric breakdown occurs. With such a design, the area which must be heated, cooled or wetted may be held to reasonable limits, allowing the anode to be immunized against breakdown and therefore achieving a high level of particle charge not possible with high-resistivity operation.

In the collecting section, the requirement for optimum collection is simply high and uniform electric field strength. High ion and current densities are *not* a specific requirement for good particle collection. Therefore, since dielectric breakdown of the high-resistivity dust layer depends on the current density through the layer, it is possible to avoid the breakdown limitation by choosing an electrode design for the collecting section which will provide high field strengths at *low* current density large-diameter discharge electrodes, for example.

Although, theoretically, no current is necessary for particle collection, a minimum current density is necessary in order to provide a holding force for the particulate on the collecting surface and thereby ensure minimum reentrainment. Also, particles that are reentrained in the collecting section must be recharged in order to be recollected. Depending on the extent of the reentrainment tendency, it may be necessary to consider adding a supplementary charging and collecting section to the two-stage design.

Ash Conditioning

Some fly ashes from coal-fired boilers are difficult to precipitate and impose serious limitations on precipitator performance levels. Chemical conditioning can correct the problems that impair particle collection for all of these applications, assuming that the basic precipitator size is adequate to meet the desired efficiency levels with ashes that behave normally.

Conditioning has been found to be effective in improving performance in every application. It has proved more cost-effective than a conventional precipitator with an increased collecting surface area. In all cases, the capital and annualized costs for conditioning an electrostatic precipitator (ESP) are lower than the comparable requirements for a conventional ESP.

There are several conditioning approaches, each of which is most appropriate in one type of application: high and low resistivity in cold-sided ESPs, and high resistivity in hot ESPs. The theory of the various conditioning mechanisms are covered in several definitive works,[15,16] and it is the intention here to discuss the more practical aspects of the art. The improvement in w_k, the migration velocity, for conditioned units is substantial and, in some cases, has been spectacular. Sulfur trioxide and ammonia are widely used as conditioning agents, while the use of alkali-metal—particularly sodium—salts is relatively new; each of these agents is discussed in some detail.

Sulfur-trioxide Conditioning

All coals contain some sulfur, so that the conditioning effects of sulfur trioxide, which occurs naturally in the combustion process, has long been known and studied. Sulfur trioxide has been the subject of a great deal of analytical work and is the conditioning agent that has been most widely adapted to field use.

Beginning with early laboratory work by Chittum and White (cited by Dismukes[16]) experimenters have shown that for temperatures below about 200°C, the primary electrical-conduction mechanism for a bulk layer of fly ash depends on surface-moisture and chemical films adsorbed on the particles. Although practitioners frequently disagree to some extent on the actual process by which surface-conducting films form, there is unanimous agreement that SO_3 combines with the particulate and greatly enhances its conductivity. Resistivity measurements and changes in operating voltage–current characteristics for ashes treated with SO_3 attest to the resistivity modifications possible with this reagent. Some secondary benefits arising from the use of SO_3 have been variously attributed to space-charge enhancement or increased cohesivity,[17] but the lowering of fly ash resistivity is clearly the primary benefit in practically all cases.

Extremely small amounts of SO_3 are required to produce large changes in the precipitation of difficult ashes; levels of 10 ppm or less are generally sufficient to lower the resistivity by about two orders of magnitude. Such levels can provide enough sulfate molecules to coat every 1-μm particle several times. The specific amount of SO_3 required to achieve a certain level of performance will depend on:

1) the sulfur content of the coal;

2) the volume and temperature of the gas;

3) the concentration and particle-size distribution of the dust;

4) the moisture content of the flue gas;

5) the size and energization of the precipitator;

6) the resistivity modification required to meet the desired performance level.

All sulfur-based conditioning methods involve the controlled distribution of a reagent from some external source into the flue gas. Some systems simply inject the reagent as a finished product—as SO_3 or H_2SO_4 gas, for example, or a specially prepared proprietary liquid chemical—into the gas system, while other processes involve conversion of some raw material into the finished product prior to injection. The most widely used process converts elemental sulfur into SO_3 gas, which is then uniformly metered into the gas stream in the desired quantity. The raw sulfur is first converted to SO_2 gas by burning in a combustion chamber, and this gas is then oxidized to SO_3 in a catalyst chamber before being injected into the flue as an air—SO_3 mixture.

Essentially, the sulfur-to-SO_3 conversion is identical to that used in the sulfuric-acid industry since the early 1900s. The sulfur is melted and the molten sulfur is completely converted to SO_2 in the burner. Efficient conversion of SO_2 to SO_3 requires a temperature in the catalyst bed of between 775 and 825°F. Some of the heat required to maintain this temperature comes from the combustion process; the remainder is supplied by an auxiliary air heater, which also supplies the necessary dilution air to control the SO_2 concentration in the catalyst bed. Conversely, the temperature downstream of the catalyst chamber must be kept high enough to prevent SO_3 condensation in the distribution system. A uniform distribution is accomplished by proper manifold and injection-nozzle design, which usually requires very high nozzle velocities. Control features typically include automatic compensation for changes in boiler load, automatic startup and shutdown of the unit and, of course, necessary instrumentation to monitor the process.

Successful applications of SO_3 conditioning span many years. There is ample evidence that the raw-material conversion processes are the most cost-effective of all those used to date. Costs associated with sulfur-burning systems are in the range of $0.04/yr for each kilowatt of generating capacity; the costs are very sensitive to the sulfur market.

The SO_3 units have had low maintenance costs when compared to the total capital value of the equipment and, equally important, there has been no evidence whatsoever of accelerated corrosion as a result of the addition of SO_3 at low temperatures[19]; this is because the concentrations involved are low.

Ammonia Conditioning

In contrast to the rationale for the use of SO_3 as a conditioning agent, ammonia has been used in situations where an excess of SO_3, produced in the combustion of high-sulfur coals, results in an ash that is overly conductive. Such ashes are prone to reentrainment, which has been shown in some cases to persist in spite of high precipitator operating voltages and large precipitator size.

The first successful application of ammonia conditioning was on a precipitator collecting catalyst fines in a petroleum-cracking process in the 1940s; the observed effect was a lowering of fly-ash resistivity. Several years later, this result was supported in work by Chittum (cited by Dismukes[16]). Investigations in the 1960s in Australia[17] and by the T.V.A. in the United States[18] were somewhat conflicting: the Australians reported an effective reduction in resistivity with the use of ammonia, whereas no large-scale resistivity change could be found in the T.V.A. study. Dismukes clarified the effects of anhydrous ammonia on U.S. coals.[16] The effects of anhydrous ammonia, when added to an ash from a high-sulfur coal, involve three distinct mechanisms: an increase in the cohesion of the ash particles, reducing the potential for reentrainment; an increase in the space charge, resulting in a higher operating potential; and modification of the resistivity. Of these, the cohesion and space-charge mechanisms predominate. The amount of ammonia required depends on the volumetric flow rate and dust concentration of the gas stream and on the required precipitator performance.

The addition of ammonia is simple. Liquid ammonia is stored in an insulated tank equipped with immersion heaters to maintain the desired tank pressure. The liquid ammonia flows to external evaporators, which operate at tank pressure. The ammonia vapor leaving the evaporators is reduced in pressure to 5 psig and mixed with dilution air supplied by a process blower. The ammonia–air mixture is transported to a distribution manifold, where it is injected into the gas stream through an array of probes. The ammonia flow is measured and controlled on the vapor side of the evaporator using an automatic control valve and a gas-flow meter.

The use of ammonia conditioning to improve precipitator performance has been gaining a status nearly equal to that of SO_3. As mentioned earlier, the Australian experience relates to a reduction in resistivity while the American experience relates to a combination of mechanisms, the least predominant of which is modification of resistivity. Further, the American experience is practically limited to ashes where the initial resistivity is especially low. Without speculating on the actual mechanisms involved, ammonia can be applied with confidence in each situation on the basis of the success it has enjoyed to date.

Ammonia-conditioning equipment requires almost no maintenance, because it is simple. Most of the operating problems reported in the literature, especially for the earliest systems, appear to be related to excessive injection levels, design and hardware quality as it relates to the method or precision of injection and, finally, the quality of the installation itself. The reported problems, which have been overcome by proper design and installation, include internal fouling of the precipitator, nozzle plugging and rate control.

Conditioning with Sodium Compounds

Certain low-sulfur western coals produce an ash which, at high ($>350°C$) precipitator operating temperatures, displays behavior similar to high-resistivity ashes in cold-side precipitators. Performance data from such installations have shown that the collecting surface area required to achieve high efficiencies in a conventional precipitator approach is so large that the capital investment or space limitations are excessive.

The positive effect of sodium on the conductivity of fly ash has been known for many years. S.R.I. has determined that the sodium in fly ash enhances both surface and bulk conduction.[19] White suggested a conditioning mechanism using an admixture of conducting particles in a continuous path with the precipitated fly-ash layer.[1] The establishment of the theorized "paths" allows the built-up negative charge on the surface of the fly-ash layer to be dissipated. With the elimination of this charge, the potential across the fly-ash layer is reduced. The result is the elimination of the formation of positive ions at the surface of the collecting surface (back corona) and the restoration of maximum precipitator efficiency.

The use of dry sodium compounds for fly-ash conditioning has resulted in the lowering of fly-ash resistivity; in laboratory evaluations the reduction has been significant. The improvement in electrical operation resulting from the elimination or reduction of the back corona has led to a doubling of the modified Deutsch migration velocity, w', in one hot-side precipitator serving a P–C boiler fuelled by low-sulphur western subbituminous coal.

The means by which the sodium is added to the fly ash depends on the individual installation. Boiler injection as a coal additive or spraying into the upper furnace has proven effective in improving precipitator operation, but the consequential increase in boiler fouling makes this an unacceptable method for many installations. Spraying of an atomized solution into the economizer outlet duct is a suitable method; it has been used in a full-scale demonstration unit since November, 1976. The addition of a pulverized soda-ash aerosol to the flue gas at the economizer outlet appears to be a more cost-effective technology. Internal flue obstructions are not a problem, as no drying time is necessary, and there is no loss in efficiency due to the addition of water, as minimal air is used to convey the aerosol to the flues.

REFERENCES

1. H. J. White, *Industrial Electrostatic Precipitation* (Addison-Wesley, Reading, Mass., 1963).
2. W. Strauss (editor), *Air Pollution Control*, Vol. 1 (Wiley-Interscience, New York, 1971) p. 227.
3. H. J. White, *Industrial Electrostatic Precipitation* (Addison-Wesley, Reading, Mass., 1963) pp. 33–181.
4. S. Oglesby, Jr., and G. B. Nichols, *A Manual of Electrostatic Precipitator Technology*, Part I (Southern Research Institute, Birmingham, Ala., 1970), pp. 23–123.
5. H. J. White, *J. Air Poll. Cont. Assoc.*, **27**, 15, 115, 206, 309 (1977).
6. E. C. Potter, *J. Air Poll. Cont. Assoc.*, **28**, 40 (1978).
7. W. T. Sproull, *J. Air Poll. Cont. Assoc.*, **15**, 50 (1965).
8. K. Ruckelshausen, *The Elimination of Dust Deposits on Technically Smooth Surfaces by Means of Rapping or Vibrating*, Ph.D. Thesis (Technischehochschule, Stuttgart, 1956).
9. H. J. Lowe and F. F. Ross, *Combustion*, 26 (October, 1977).
10. D. Juricic and G. Herman, A.S.M.E. Trans. J. Mach. Design, Paper 77-DET-98 (1977).
11. F. S. Jeselsohn and G. M. Hieber, *Rapping Experiments on Models of Precipitator Collecting Surfaces* (Electric Power Research Institute, to be published).
12. O. J. Tassicker, *Aspects of Forces on Charged Particles in Electrostatic Precipitators*, Ph.D. Thesis (University of New South Wales, Australia, 1972).
13. H. W. Spencer, III, *Rapping Reentrainment in Nearly Full Scale Pilot ESP* (Southern Research Institute, Birmingham, Ala., 1976).
14. P. L. Feldman, *Effects of Particle Size Distribution on Performance of Electrostatic Precipitators* (APCA Mtg., Boston, Mass., 1975).
15. R. E. Bickelhaupt, *Effect of Chemical Composition on Surface Resistivity*, NTIS PB 244885 (August, 1975).
16. E. B. Dismukes, *Conditioning of Fly Ash with* SO_3 *and* NH_3, NTIS PB 247231 (1975).
17. K. S. Watson and K. J. Blecker, *Int. J. Air Water Poll.*, **10**, 573 (1966).
18. J. T. Reese and J. Greco, *J. Air Poll. Cont. Assoc.*, **18**, 523 (1968).
19. R. E. Bikelhaupt, *Sodium Conditioning to Reduce Fly Ash Resistivity* EPA-650/2-74-092 (1974).

Peter B. Lederman
and Paul L. Feldman
Research Cottrell
Bound Brook, New Jersey

INDEX

Acid brown earth 68
Acidity 56
Acid precipitation 56
Activated carbon 17, 19
Advective integral model 186
Aerobic 107
Aerobes 80
Algae 65
Altitude—latitude model 130
Amazon Basin 61
Ames, Iowa 21
Amino acids 64
Ammonia conditioning 218
Ammonia stripping 17
Anthropogenic 60, 101, 151, 171
Annual precipitation 57
Antioxidant 84
Ash 66
Ash conditioning 216
Atlantic Ocean 107

Backstrom mechanism 82
B.A.T. see Best Available Technology Economically Achievable
B.C.T. see Best Conventional Pollutant Control Technology
Best Available Technology Currently Available 1
Best Conventional Pollutant Control Technology 4
Best Management Practices 9
Best Practicable Control Technology 1
Biochemical Oxygen Demand 4, 14
Biocirculation system 64
Biogenic production 110
Bilogical ion exchange 62
Biological treatment 14
Bioresistant materials 24
Biosynthesis 64
Birkenes, Norway 75
B.O.D. see Biochemical Oxygen Demand
Boundary layer 184
B.P.T. see Best Practicable Control Technology Currently Available
Budgets 135, 144, 171
Buoyant jets 34

Canopy layer 183
Carbon monoxide 100
Chemical Oxygen Demand 16
Chlorination 17, 19, 25
Clean Water Act 1
Climatology 171
Coagulation 16, 19
Coherence 51
Cold islands 177
Coliforms 4, 15
Colorado State University 192
Combustion 114, 129
Comfort 171
Compatibility 23
Complexing 82
Consent decreee 3
Construction grants 6
Conventional pollutants 4
Cooling systems 31
Cooling towers 32
Cooling water 30
Corps of Engineers 12
Crossover layer 178
Cost relationships 19
Cost sharing 10
Counter ion 81
Coupling 137

DaVinci flight 166
Desorption 99
Deutsch–Anderson equation 210
Diazotization 93
Diffusion 35, 98, 116
Diorites 60
Disinfection 17
Diurnal variation 154, 175
Downslope flow 155
Dust 206

Economic analysis 19
Ecosystem 71

Eddies 49
Eddy diffusion 116, 185
Effluent—limitation guidelines 10
Electrostatic precipitators 203
Emmissivity 174
Energy balance 184
Energy requirements 19, 22
Entrainment 46
Environmental impact 19
Environmental Protection Agency 3, 7
E.P.A. see Environmental Protection Agency
Evapotranspiration 61
Exchange capacity 69

Fabric filter 204
Federal Water Pollution Control Act 1
Feldspar 65
Field charging 208
Filtration 17, 19
Fishable/swimmable 1
Fish and Wildlife Service 12
Fish mortality 58
Fixed solids 15
Flocculation 16
Flow equalization 24
Flux 108
Fog 86, 181
Forest fires 151
Forest vegetation 56
Free radical 118
Fresh water 56
Froude number 34, 43
Fuel inventories 177
Fungi 112
F.W.P.C.A. see Federal Water Pollution Control Act

Geochemistry 71
Grass 69

Health 171
Heat 3
Heather 71
Heat island 178
Heat wave 172
Henry's Law 93, 95
Heterogeneous 128
High pressure system 164
Horizons 63
Hot precipitators 212
Humic acids 59
Humidity 181
Humus 69
Hydrochloric acid 57

Industrial wastewaters 14
Inhibition 80
Innovative technology 7
Interdisciplinary 143
Intermittency 49, 53
International Atomic Energy Agency 30
Ion exchange 26, 58
Irrigation return flows 11

Kaolin 61

Lagrangian sampling 166
Lake Tahoe 22
Litterfall 74
Logging 67
Log—normal 210
Los Angeles 102

Mannitol 83
Marine waters 5
Mass transfer 92, 108
Max Planck Institut 106
Methane 100
M.E.T.R.O.M.E.X. see Metropolitan Meteorological Experiment
Metropolitan Meteorological Experiment 182
Milk 70
Mixing 33
Mixing height 190
Mixing layer 154
Monitoring 139
Moisture balance 176

N.A.A.Q.S. see National Ambient Air Quality Standard
National Ambient Air Quality Standard 151
National Pollutant Discharge Elimination System 2
Naval Research Laboratory 106
New York–New Jersey area 162
Nitric acid 92
Nitrification 68
Nitrification—denitrification 19
Nitrogen 15, 25
Nitrogen dioxide 92
Non conventional pollutants 4
Non point sources 9
Norwegian Acid Precipitation Project 56, 75, 77
N.P.D.E.S. see National Pollutant Discharge Elimination System

Numerical modeling 36
Nutrients 15
Nutrient removal 25

Oceans 105, 107
Oxidation 80
Ozone 80, 125, 150
Ozonation 17, 19, 27

Pacific Ocean 105
Partial pressure 63, 80, 95
Penetration theory 95
Perturbation 127
pH 4, 57, 71
Phenolic 87
Phosphorous 15, 25
Photochemical 102, 119
Photochemistry 137
Photolytic 101
Physical–chemical processes 14
Planetary boundary layer 178
Plume 34, 88, 151, 160
Point source 1, 5
Political trade-off 9
Pollen 71
Podzol profile 58
Power plants 30
Precipitation 177
Proton pumping 59
Pulsed energization 214

Quartz 65

Radiation inversion 153
Rain 86
Rainfall intensity 75
Rainwater 57
Rapping 204
Reentrainment 206
Research Tringle Institute 157
Reliability 21
Residue 27
Resistivity 203
Reynolds number 34, 42
Reynolds stress 52
Runge–Kutta approximation 94
Runoff 69
Rural set—aside 7

Saturated hydrocarbons 158
Scale models 191
Schönbein method 150
Secondary treatment 5
Secretary of Agriculture 10
Section 404 11
Short wave radiation 173
Simulation 134
Sink 102, 107, 113
Skin cancer 173
Sludge 8, 17, 28
Smelters 57
Smog 158
Smog-chamber 159
Snow 65
S.N.S.F. see Norwegian Acid Precipitation Project
Soil 111
Soil Conservation Service 10
Soil erosion 56
Solar energy 31
Source 102, 105, 111, 138
Sphagnum peat 62
Stanford Research Institute 156
Stratosphere 115
Subsidence inversion 160
Sudbury, Canada 56
Sulfate 57
Sulfide 63
Sulfite 88
Sulfur dioxide 80
Sulfuric acid 57
Sulfur trioxide 216
Supersaturation 107
Suspended solids 4, 14
Synergistic effects 83

Temperature 33, 112, 137, 177
Tertiary treatment 14
Thermal effects 31
Thermal wastewater 53
Three-dimensional model 132
Topsoil 64
Toxic pollutants 3, 102
Trace gases 137
Trace species 100
Transport 117
Tropopause 115
Troposphere 100
Turbidity 27
Turbulence 44, 48, 133, 209
Two-stage precipitator 213

Uptake exchange coefficient 113
Uptake rate 111
Urban climate 171
User charges 8
U.S. Standard Atmosphere 124

Vertical profiles 107
Visibility 182
Volatile suspended solids 15

Wastewater 14
Water 56
Water droplets 92
Water vapor 98
Weathering 60
Weather modification 171
Wet precipitation 213
Wind 179
World Meteorological Organization 172

Zero discharge 3